MODERN ELECTRONICS GUIDEBOOK

An Overview

Michael N. Kozicki
Arizona State University

VNR VAN NOSTRAND REINHOLD
New York

Library of Congress Catalog Card Number 91-26192
ISBN 0-442-00612-8

Manufactured in the United States of America

Published by Van Nostrand Reinhold
115 Fifth Avenue
New York, New York 10003

Chapman and Hall
2-6 Boundary Row
London, SE1 8HN, England

Thomas Nelson Australia
102 Dodds Street
South Melbourne 3205
Victoria, Australia

Nelson Canada
1120 Birchmount Road
Scarborough, Ontario MIK 5G4, Canada

16 15 14 13 12 11 10 9 8 7 6 5 4 3 2 1

Library of Congress Cataloging-in-Publication Data

    Kozicki , M. (Michael), 1958-
    Modern electronics guidebook: an overview/Michael N. Kozicki.
        p. cm.
    Includes bibliographical references and index.
    ISBN 0-442-00612-8
    1. Electronics--Handbooks, manuals, etc. I. Title.
    TK7825.K69 1991   91-26192
        621.381--dc20   CIP

for Matthew and Anna

Contents

Preface

As members of a high-technology society, we find ourselves surrounded by the fruits of microelectronics. The new technology has made possible a second industrial revolution, a golden age of manufacturing which has provided the population with fantastic computing and communications power, now central in industry and commerce, as well as in our leisure and home life. We find integrated circuits, "silicon chips", giving capability and flexibility to a vast range of goods and systems from the modern washer in your laundry, to the communications satellite far above in Earth orbit handling your trans-pacific telephone call. Items such as the "palm-top" computer and the "wrist watch pager" are testament to the ability of the technology to pack a great many functions into a small volume at reasonable cost and with excellent reliability. Indeed, it is these four factors (capability, size, cost, reliability) which make the items produced by microelectronics technology so desirable in today's society. However, it is a somewhat disturbing fact that even though we are all touched by the technology, very few of us know and understand even the most basic facts relating to microelectronics.

As the title suggests, this book is a guide through the jargon jungle of microelectronics. It is something of a "crossover" text in that its content and style allows it to be understood and appreciated by people on a technical career path (for instance, electrical engineering/technology students), as well as others who have to deal with the technology in some respect on a day-to-day basis. It is structured in such a way as to be of considerable value to executives and management in high-technology firms and will allow them to become conversant with the terms and concepts used by their technical colleagues.

The book begins with a light history of the technology to put it in perspective and then takes the reader through the basic physics of electronic materials, describes the properties of the materials themselves, discusses basic device action, and shows how devices are actually made. The bulk of the device discussion is centered around transistors as these are the most important building blocks in modern circuits, whether the circuits are built up on a printed wiring board or are made in the form of an ultra-modern silicon chip. It also discusses the operation of a wide range of different device types, including their particular merits and limitations, from the elements which make up the memory circuits in the personal computer to the units which form the imaging system in the ubiquitous camcorder. The book also gives examples of circuits containing the most common electronic devices to illustrate their applications.

The book is undeniably technical, ranging from rudimentary concepts to the state of the art in materials and devices, but is nevertheless highly readable. It presents complex concepts without the usual mathematical bombardment to increase its appeal to all who have an interest in the technology. It will thus give engineering and applied science students a good foundation for more advanced courses which do have a mathematical element but it is also sufficiently complete and self-contained so that it may be used in technology awareness courses and the like in business colleges and corporations.

This book was made possible by the teaching and influence of a great many people over the years, too many to thank individually here, but I am particularly grateful for the invaluable assistance of Susan Alexander in the production of the final version of the text.

Glossary

This short glossary contains definitions of many key words and phrases used in modern electronics. More are defined within the text and may be found by using the Index at the end of the book. Important terms are italicized the first time they appear in the text.

Accelerated testing—increased temperature may be used to speed-up potential failure modes in a sample component so a lifetime of stress is experienced in a few hours. Some idea about component reliability can be determined using this type of test.

Acceptors—deliberately introduced impurity atoms which grab an electron from the semiconductor to allow the formation of a positively charged mobile "hole".

Active devices—electrical components, such as transistors, able to perform amplification functions.

Address bus—the wiring and associated switches in a computer system which carry, in digital form, information regarding the location in the computer memory of data or program information.

Amplification—a small voltage or current used to control a larger voltage or current.

Analog—a continuous variable, able to take on any value within a range.

Analog-to-digital converters (ADC)—circuits which can take a continuous (real world) signal and convert it to digital or numerical (computer) form.

Application-specific integrated circuits (ASICs)—ICs not functionally generic, but designed for one specific application.

Arithmetic and logic unit (ALU)—circuit within a microprocessor which performs arithmetic or logical operations on information in digital form.

Baud rate—the rate (in bits per second) of digital information transfer in a circuit connecting computers together.

Binary—able to take only two values, 0 or 1 (see also logic states).

Bipolar junction transistor (BJT)—a transistor, consisting of two back-to-back diodes, which uses both electrons and holes in its operation.

Bit—the smallest unit of digital information, binary in nature.

Boolean algebra—a form of mathematics used in logical operations.

Burn-in—all components are made to operate in a high temperature environment to weed-out those which were destined to fail early (the "infant mortality" effect), so that they are not passed on to the customer.

Byte—a group of eight bits.

Charge carriers—mobile entities which carry charge in materials; electrons carry negative charge and holes carry positive charge.

Charge-coupled device (CCD)—device in which packets of electric charge may be transferred along its length by sequentially pulsing its "gate" electrodes.

Chemical vapor deposition (CVD)—a thermal method used to form thin layers of metals, semiconductors and insulators in the manufacture of ICs.

Chip—the common name for an integrated circuit.

Cleanroom—the controlled environment, largely free of dust and other contamination, used in the manufacture of ICs.

Complementary MOS—the combination of NMOS and PMOS devices in the same IC.

Compound semiconductors—semiconducting materials which consist of more than one element, e.g. gallium arsenide (Ga + As).

Conductors—materials which have many charge carriers and can, therefore, conduct an electric current well.

Crystal puller—machine used to produce near-perfect semiconductor crystals which act as substrates for ICs.

Data bus—wiring and associated switches which carry information in digital form in a computer system.

Depletion-mode MOSFET—MOS transistor which requires a voltage on its gate electrode to switch it fully off.

Dielectrics—materials in which the charge carriers are tied up in the bonds between atoms and are, therefore, poor conductors of electric current.

Diffusion—thermal process used to introduce dopants into semiconductors.

Digital—pertaining to numbers or a variable represented by discrete numerical values.

Digital-to-analog converters (DACs)—circuits which can take a numerical variable and convert it to an analog signal.

Diode—electronic component which only allows substantial current flow in one direction.

Donors—deliberately introduced impurity atoms which give-up an electron to the semiconductor to provide more negatively charged carriers.

Dopants—acceptor or donor atoms in a semiconductor.

Drop-ins—test circuits on a wafer processed along with the product circuits.

Dual in-line package (DIL or DIP)—a protective package for ICs with connections to external circuits arranged along the long edges of the package.

Dynamic random access memories (DRAMs)—electronic memories which store information in digital form. Information may be accessed at any point in the memory, but all information has to be rewritten (refreshed) regularly.

Electrically erasable PROM (EEPROM)—a PROM which may be erased using an electrical signal rather than by exposure to ultra-violet light.

Electrons—the outer layers of atoms which take part in chemical reactions, such as bonding, and the carriers of negative charge in materials.

Elemental semiconductors—semiconductors which consist of only one element, e.g. silicon (Si).

Enhancement-mode MOSFET—MOS transistor which is off when no voltage is applied to its gate electrode.

Erasable PROM (EPROM)—PROM which can be erased (information removed) using ultra-violet light irradiation.

Etching—process using wet chemicals or gas plasmas for removing regions from the surface of an IC during production.

Evaporation—thermal process in which thin metal films may be deposited on ICs during manufacture.

Field effect transistor (FET)—transistor in which an electric field on a gate electrode is used to control the current passing through the device.

Full custom design—design process in which few predesigned elements are used and each part of the circuit is designed to attain maximum performance and packing density.

Functional test—test performed at the end of processing to determine whether the component meets all functional requirements.

Gallium arsenide—common compound semiconductor typically favored in applications which demand high speed/high frequency operation.

Gate—control electrode in a field effect transistor, or a circuit which performs a logical operation.

Gate arrays—collection of pre-processed logic gates on a substrate which may be wired by patterning the final levels of metallization to produce a particular digital circuit.

Heterojunction bipolar transistor (HBT)—bipolar transistor consisting of different (usually compound) semiconductors.

Hybrid—a circuit, based on a ceramic substrate, which combines thick-film (screen printed) resistors and chip capacitors with ICs.

Integrated circuit (IC)—circuit which has been fabricated on a semiconductor substrate.

Ion—atom which has either lost or gained one or more electrons to become net positively or negatively charged respectively.

Ion implantation—particle acceleration technique used to introduce highly controlled amounts of dopants into semiconductors during manufacture.

Light-emitting diodes (LEDs)—diodes (p-n junctions) made from compound semiconductors which can emit light for display or communications purposes.

Logic states—two possible logical values a binary variable can take, false or true, logic 0 or logic 1.

Logical operations—mathematical manipulation of logical variables (see also Boolean algebra).

Mainframes—large, non-portable, multi-user computers.

Mask aligner—an optical system used in the manufacture of ICs to form and align the patterns on the substrate which make up the devices and circuits. Light is projected through a mask in the image transfer process.

Mean time between failures (MTBF)—average expected time a component will last before failing.

Memory—circuit used to store information in digital form.

Metal-nitride-oxide-semiconductor (MNOS)—MOS transistor which can act as a nonvolatile memory by storing charge between the oxide and nitride layers.

Microprocessor—IC which contains the necessary circuitry to carry out instructions governed by a stored program and perform arithmetic and logical operations on data in digital form.

Monolithic microwave integrated circuits (MMICs)—ICs which can operate at frequencies in the thousands of millions of cycles per second regime.

Nonvolatile memory (NVM)—memory which will retain digital information even when the power is removed from the circuit.

Operational amplifier—analog circuit which can amplify a small signal and which is also a fundamental building block in many different analog systems.

Oxidation—thermal process in which a layer of silicon dioxide is grown on a silicon substrate to form, for example, the gate dielectrics in MOS devices.

Parametric test—testing which is performed, usually during processing, to determine whether certain fundamental parameters are within specification.

Passive devices—circuit elements, such as resistors, which cannot perform amplification functions alone.

Photolithography—process in which a pattern on a mask is transferred to a photosensitive layer on a substrate for the purpose of delineating the layers of a device/circuit.

Planar technology—technique in which processing operations are performed on the surface of a semiconductor substrate to build-up an IC.

Power devices—electronic components which can handle high currents or voltages.

Printed wiring boards (PWBs)—circuit boards used to connect ICs and discrete components together.

Program—sequence of instructions a computer follows, stored in memory.

Programmable logic arrays (PLAs)—see gate arrays.

Programmable read-only memory (PROM)—ROM programmed by the user (it does not have to be programmed at the manufacturers).

Quantum mechanics—physics which describes how very small entities, like electrons, behave.

Radiation hard—device or circuit which is tolerant to the effects of ionizing radiation.

Random access memory (RAM)—electronic memory in which any address location may be written to or read from when desired.

Read-only memory (ROM)—electronic memory which has fixed data which can be read at any time but which is generally programmed in only once.

Semiconductors—class of materials which are good insulators at very low temperature, good conductors at very high temperature, but are somewhere in between at room temperature. The introduction of dopant impurities alters the electrical characteristics to allow the formation of devices.

Silicon—most commonly used elemental semiconductor due to its abundance and ease of processing.

Silicon-controlled rectifiers (SCRs)—semiconductor devices used to control ac power (e.g., in light dimmers or electric motor speed control).

Small signal—devices and circuits which handle only small currents and voltages.

Sputtering—method used in the fabrication of ICs in which conducting or insulating material is ejected from a target by ion bombardment to condense as a thin film on a substrate.

Standard cell—circuit design technique in which the circuit is built-up using pre-designed sections from a cell library.

Static RAM (SRAM)—information is stored in electrical form as in the case of a DRAM but this type of RAM does not require the information to be continually rewritten to it.

Superconductors—class of material in which the resistance to current flow drops to zero below a critical temperature.

Surface mount technology (SMT)—packages connected onto the surface of a wiring board, i.e., the legs of the packages do not protrude through the board as in normal PWBs.

Transistor—semiconductor device in which a small voltage or current can be used to control a larger voltage or current. It may therefore act as an amplifier or a switch.

Uncommitted logic arrays (ULAs)—see gate arrays.

Wafers—thin semiconductor substrates, cut from a larger crystal, on which the ICs are built-up.

Words—groups of bytes.

Yield—number of working circuits expressed as a percentage of the total number manufactured.

Abbreviations

ac	alternating current
ADC	analog-to-digital converter
ALU	arithmetic and logic unit
APCVD	atmospheric pressure chemical vapor deposition
ASIC	application specific integrated circuit
ATE	automatic test equipment
BiCMOS	bipolar-CMOS
BJT	bipolar junction transistor
CD	critical dimension
CCD	charge-coupled device
CMOS	complementary metal-oxide semiconductor
CRT	cathode ray tube
CVD	chemical vapor deposition
CZ	czochralski
DAC	digital-to-analog converter
dc	direct current
DI	dielectric isolation or deionized (water)
DIP	dual in-line package
DRAM	dynamic random access memory
DSB	direct silicon bonding
EAROM	electrically alterable read only memory
FAMOS	floating gate MOS
FET	field effect transistor
FZ	float zone
GTO	gate turn-off thyristor
HBT	heterojunction bipolar transistor
HEMT	high electron mobility transistor
HEPA	high efficiency particulate air (filter)
HEXFET	hexagonal field effect transistor
HMOS	high-performance metal-oxide semiconductor

IC	integrated circuit
IGFET	insulated gate field effect transistor
IR	infra red
I^{2}L	integrated injection logic
ILD	interlayer dielectric
JFET	junction field effect transistor
LCD	liquid crystal display
LEC	liquid encapsulated czochralski
LED	light-emitting diode
LDD	lightly doped drain
LPCVD	low pressure chemical vapor deposition
LSI	large-scale integration
MBE	molecular beam epitaxy
MESFET	metal-semiconductor field effect transistor
MIPs	millions of instructions (executed) per second
MISFET	metal-insulator-semiconductor field effect transistor
MNOS	metal-nitride-oxide-semiconductor
MOSFET	metal-oxide-semiconductor field effect transistor
MOS	metal-oxide-semiconductor
MSI	medium-scale integration
MTBF	mean time between failure
NMOS	n-channel MOS
NVM	nonvolatile memory
PGA	pin grid array
PLA	programmable logic array
PMOS	p-channel MOS
poly-Si	polysilicon
ppm	part per million
PROM	programmable read only memory
PSG	phosphosilicate glass
PVD	physical vapor deposition
PWB	printed wiring board
QA	quality assurance
QC	quality control
RAM	random access memory
rf	radio frequency
RIE	reactive ion etching
ROM	read-only memory
RTP	rapid thermal processing
SAG	self aligned gate
SAW	surface acoustic wave

SCR	silicon-controlled rectifier
SOI	silicon on insulator
SOS	silicon on sapphire
SRAM	static random access memory
SSI	small-scale integration
ULA	uncommitted logic array
ULSI	ultra-large-scale integration
VHSIC	very high speed integrated circuit
VLSI	very-large-scale integration
VFET	vertical field effect transistor
WSI	wafer scale integration

1

Electronics— Vacuum Tubes to Computers

The Path to Modern Electronics

It is difficult to say where the path which ultimately led to the electronic components of the information age actually began. If we wish, we could go back to ancient times to find the roots of electronics technology in the discovery of magnetism. However, it probably makes more sense to start at the point where the first rudimentary electronic components appeared and to highlight the major milestones on the way to the electronic devices, circuits, and systems of the present.

Fleming (1904)—invents the vacuum tube *diode*, a key *device* in electronics. The distinguishing characteristic of the diode is that it allows electricity to flow in one direction but prevents flow in the reverse direction. Diodes are used in communications, information processing, and power conversion.

The original diodes were based on vacuum tube technology in which the electric current, which is merely a flow of electrons, passed through a vacuum between electrodes inside a glass envelope. A vacuum is necessary, as the presence of gas would interfere with the electron flow. One electrode, the *cathode*, is heated to eject the electrons from the metal (a process known as *thermionic emission*). After traveling through the vacuum, the electrons are collected at the other electrode, the *anode* or *plate*. If the plate is connected to a positive voltage with respect to the cathode, the negatively charged electrons are attracted to it (remember, unlike charges attract) and an electrical current passes through the device. If we make the cathode positive and the plate negative (but still release electrons by heating the cathode), the electrons cannot travel across the vacuum, as they are attracted back to the positively charged cathode and hence stay nearby. Therefore, current can flow in only one direction in the device; this phenomenon is known as *rectification*.

In addition, the relationship between the size of the voltage between the anode and cathode and how much current flows in the device is *nonlinear*; as the voltage is increased, the amount of current which flows increasing approximately as the square of the voltage. Diodes are thus used in communications as *detectors*—elements which separate the *signal* (voice, music, etc.) from the *carrier* (the high frequency which we actually tune our radios to). Detection requires nonlinear device characteristics (this can be shown mathematically but is beyond the scope of this book).

de Forest (1906)—invents the *triode amplifier*, a device which allows current flow to be controlled. This became an important device in practically all areas of electronics but was originally applied to the *amplification* (strengthening) of electrical signals. The triode was the predecessor of the *transistor*.

If we take our rectifier and place a mesh-like third electrode, the *grid*, in the path of the traveling electrons, we can turn some or all of the electrons away from the anode. The amount of current held back in this way depends on the size of the negative voltage we apply to the grid (like charges repel; hence the negatively charged grid will tend to push the negatively charged electrons away). We have thus created a triode amplifier in which a small voltage change on the grid can control a large current swing between cathode and anode (Fig. 1.1). This device could also be used as an electronic *switch*, as a large change in grid voltage could be used to turn the flow of current on or off. Incidentally, if the anode is replaced by a screen coated with a material which will give off light when struck by the traveling electrons and the electron flow is diverted by charged electrodes

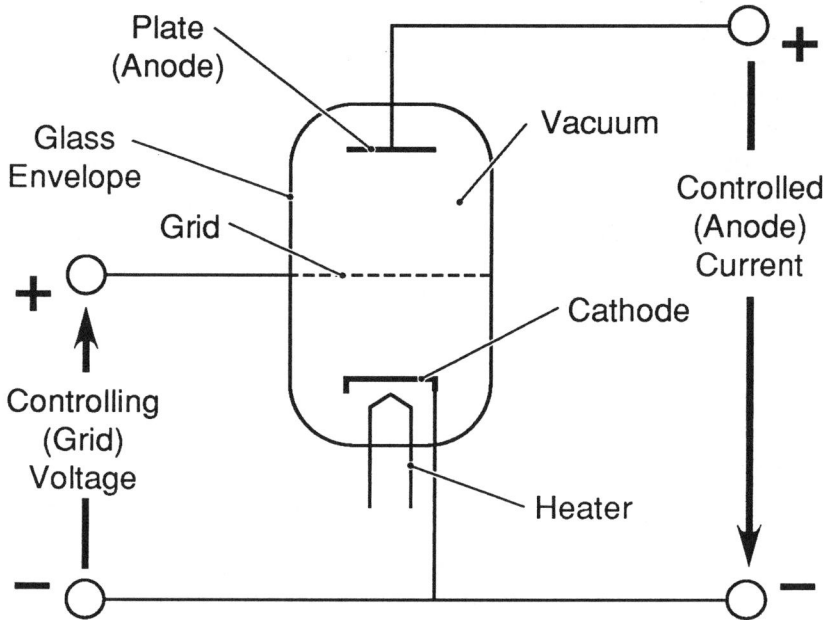

Figure 1.1 Schematic diagram of the vacuum tube triode.

to scan across this screen, we effectively have created a *cathode ray tube* (CRT) or television tube.

Armstrong (1912)—develops the electronic *oscillator*. An oscillator creates an electronic signal of a particular frequency (to use a familiar analogy, it is like an electronic tuning fork). The first oscillators utilized basic vacuum tube components in a *circuit* (a connected set of devices) which would find many uses in the future, particularly in communications.

World War II (1939-1945)—forces the miniaturization of electronics. The war provided the thrust for many technological developments, including the portability of many electronic systems. This meant that the components which made up the circuits (vacuum tubes, *resistors, capacitors* and *coils/inductors/transformers*—see Chapter 3), had to become smaller and lighter and consume less power.

Eckert and Mauchly (IBM, 1946)—develop the *vacuum tube computer*. This was an astounding feat of design and construction and was the first truly electronic computer to be developed. It utilized the switching capabilities of the vacuum tubes to perform arithmetic operations (see later). Due to the inherent complexity and unreliability of the vacuum tube approach, few systems of this type were produced. Vacuum tubes tended to be unreliable, as the cathode heaters burned out in much the same way as a light bulb filament. They were also very fragile and easily failed if handled roughly. Their tendency to fail meant that they had to be connected into circuits by means of sockets, and this also produced more reliability problems as well as increasing the overall cost.

Shockley, Bardeen and Brattain (Bell Labs, 1947)—invent the *transistor*. This was the three-terminal component which would ultimately replace the vacuum tube in all but a few applications. The device was based on *solid-state* technology involving *semiconductors* rather than relying on heaters or fragile evacuated envelopes. Solid-state devices were thus considerably more reliable than their vacuum tube counterparts, and could be made much smaller and more cheaply.

IBM (1955)—produces the 7090 *transistorized computer*. This represented the pinnacle of achievement in the use of solid-state components in a major electronic system (an interconnected set of circuits). The inherent reliability of transistors meant that the components could be soldered in place, which facilitated the task of interconnecting them. The overall system was also much smaller and, since it contained more components, more powerful than its vacuum tube counterparts.

Hoerni (Fairchild, 1958)—suggests the concept of *planar technology*. Up to this point, the formation of transistors was considerably more simple than the making of vacuum tubes, but it was still a tricky process in terms of handling the materials and controlling how well the devices operated. In planar technology, all the processes used to make the devices are performed on one side of a flat piece of semiconductor. This considerably simplifies the production process and opens up other possibilities, which will be discussed in greater detail later in this text.

Noyce (Fairchild, 1958)—proposes the concept of *integration*. This idea utilized the principles of planar technology to create not one device on the flat piece of semiconductor but many devices simultaneously. This allowed entire circuits to be formed on a semiconductor *chip* to create a *monolithic* or *integrated circuit (IC)*.

Kilby (Texas Instruments, 1958)—develops the first IC. This did not exactly follow the Noyce concept, as it was developed separately, but it did represent the first real multicomponent IC.

1960-1962—ICs come to market. The technology of the IC was quick to mature. Commercially available units appeared within four years of the first relatively crude prototypes.

Hoff (1969)—develops the integrated *microprocessor*. The micro-processor is effectively an integrated computer with *program storage (memory) and control*, an *arithmetic and logic unit (ALU)*, and a small amount of *scratch-pad memory* on one chip of *silicon*. Silicon rapidly became the semiconductor material of choice due to its low cost and ease of processing. Note: As testament to the complexities of the patent process, although Hoff's name appears in the history books as the developer of the microprocessor, the patent for the single-chip computer was actually awarded to **Hyatt** in 1990!

Intel (1971)—produces a commercial *integrated computer*. Utilizing the microprocessor concept and adding *program* and *data* memory, Intel developed an off-the-shelf *microcomputer*.

It must be appreciated that in the early days of electronics, all devices were *discrete*, existing as individual units within their own packages. Vacuum tubes, transistors, and other components were separate entities and had to be joined electrically by wires to other devices, including simple elements such as resistors and capacitors. The work of Kilby and Noyce in 1958 gave rise to IC technology, which made it possible to create many devices arranged together, or integrated, as an electrical circuit on a single piece or chip of semiconducting material. As the technology matured, increasing numbers of devices could be integrated to form increasingly complex circuits (Fig. 1.2). From the early days when a few tens of components were integrated on a chip, the trend has continued relentlessly, so that we are now witnessing the development of integrated circuits which contain more than 64 million transistors.

There are a number of definitions which relate to integration. These are (with the approximate time that they appeared commercially):

Small-scale integration (SSI)—early 1960s. A few tens of components per circuit.

Medium-scale integration (MSI)—mid-1960s. Several hundred components per circuit.

Large-scale integration (LSI)—1970s. A few thousand components per circuit.

Very large-scale integration (VLSI)—1980s. Tens of thousands to over 1 million components per circuit.

Ultra large-scale integration (ULSI)—1990s. Several million to tens of millions of components per circuit.

We may also add *very-high-speed integrated circuits (VHSIC)* as an additional definition. This military program appeared in the late 1980s to create extremely fast-operating VLSI/ULSI circuits.

The Drive for Smaller Devices

As we will see in Chapter 3, larger-scale integration produces circuits which are smaller, faster, more reliable, and cheaper. How do we achieve higher levels of integration? We make the devices which make up the circuit smaller and pack them together more densely.

In addition to the number of components in a circuit, a measure of the scale of integration is the *linewidth*. This is the smallest defined dimension within the circuit, for example, the width of a conductor or element within a device, usually measured in *microns* (1 micron = 1 millionth of a meter). To put this unit into perspective, a human hair is on the order of 80 microns in diameter. In the original discrete devices, linewidth could be thousands of microns; SSI, several hundred microns; MSI, several tens of microns; LSI, 3 to 10 microns; VLSI, 1 to 3 microns; and ULSI, less than 1 micron. It is important to

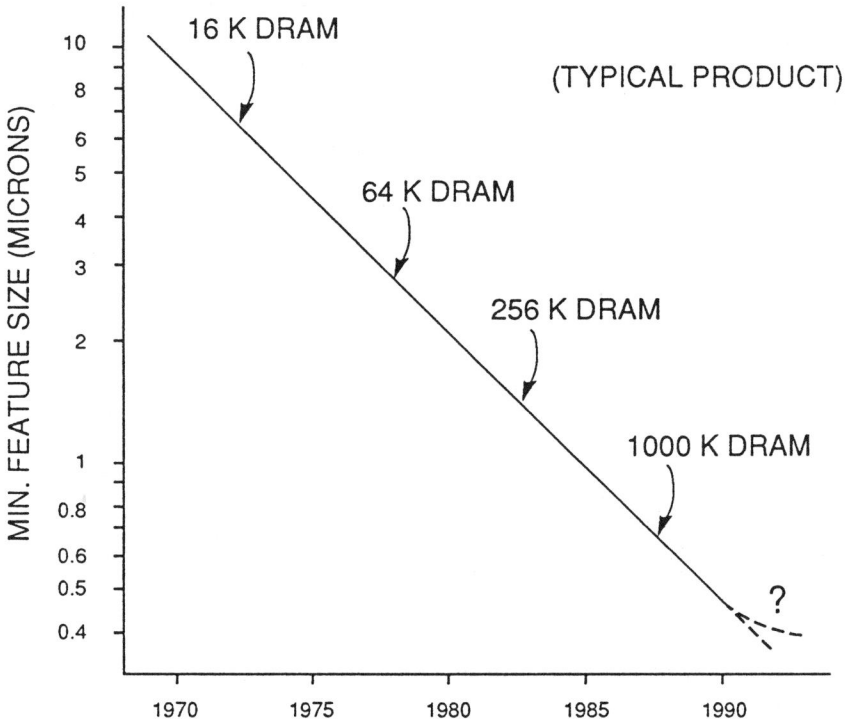

Figure 1.2 Minimum IC feature size vs. year.

understand that higher levels of integration demand smaller linewidths. Smaller linewidths come with a very high price, as they may be achieved only through a greater understanding and control of materials and fabrication technology. This is the subject of a substantial portion of this book. However, higher levels of integration allow circuits to be produced that would be impossible with any other technology due to cost and reliability considerations. Without integration, we would not be in the information age.

The Real World and Numbers

Electrical circuits and *systems* (collections of circuits) fall into two broad categories—those that handle *analog* quantities and those that handle *digital* variables. Analog quantities are *continuous* (Fig. 1.3a). The natural world is an analog world; variables such as height, temperature, and weight are continuous in that they vary smoothly. Through the use of an appropriate *transducer*, a device which can translate real-world variables into *electrical signals* (voltages and

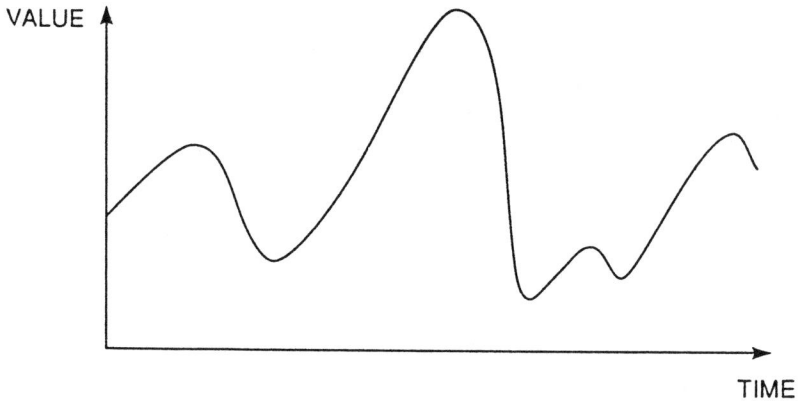

(a) Analog variable

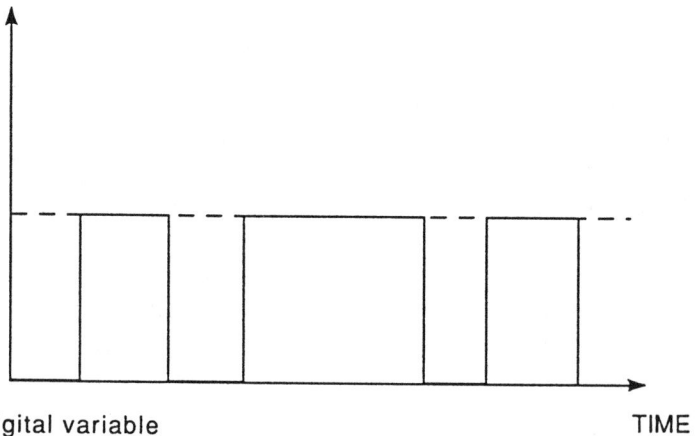

(b) Digital variable

Figure 1.3 Examples of (a) analog and (b) digital variables showing the discrete levels in the digital case.

currents), analog circuits can act on or *process* information from the natural world. An example of an analog circuit is a public address system in which the analog variable, the human voice, is transformed into an electrical signal by a transducer—a microphone. A circuit within the system then boosts the electrical signal by transistor amplification techniques and then puts out a louder version of the voice by way of an electrical signal-to-sound transducer—a loudspeaker. Analog circuits contain *active* (able to perform amplification) devices such as transistors and also passive components such as resistors, which may be used to reduce current or divide voltage, capacitors, which store charge and are also used in *tuning circuits* and oscillators, and inductors, which are used in voltage conversion units called *transformers* and tuning circuits.

Digital variables are in the form of *discrete levels* with no intermediate values (Fig. 1.3b). These levels may be represented by numbers, which in turn are represented by the *logic states*, *logic 0* and *logic 1*. Logic 0, in electrical terms, could correspond to the off position or off state of a switch, or, alternatively, may just mean the absence of a voltage or current at a particular point in a circuit. Logic 1 would then correspond to the on state, or the presence of a particular voltage or current.

Due to their *binary* nature—that is, having two levels corresponding to the numbers 0 and 1—the logic states are used to represent numbers in *base two* which have equivalents in our usual counting scheme, *base ten* (see Table 1.1). For example, 0 in base two = 0 in base ten and 1 in base two = 1 in base ten, 10 (said "one-zero") in base two = 2 in base ten, 11 ("one-one") = 3, 100 = 4, 101101 = 45, etc. In *computing* terms, each binary digit in the base two number is called a *bit*, an 8-bit group (e.g., 00000010, 10010110) is called a *byte*, and groups of bytes are generally called *words*. To save writing all the zeros and ones, the byte or word values are usually represented using the *hexadecimal* system. Here we take groups of four bits out of a word, which can have a maximum value of 15 (1111 = 15), and represent them with their hexadecimal value. In the hexadecimal system, zero—nine have their normal base ten meaning but ten is represented by A, 11 by B, 12 by C, 13 by D, 14 by E, and 15 by F. The binary number 1100 may thus be written as C, 10001100 as 8C (note how the first group of four, 1000, equals 8, and the second group, 1100, equals 12 or C), 1101001010001100 as D28C, and so on. This form of representation of binary numbers is frequently used for 16-bit data values and *address* (where the data is stored) locations.

	$2^5 = 32$		$2^4 = 16$		$2^3 = 8$		$2^2 = 4$		$2^1 = 2$		$2^0 = 1$
0 =	0	+	0	+	0	+	0	+	0	+	0
1 =	0	+	0	+	0	+	0	+	0	+	1
2 =	0	+	0	+	0	+	0	+	1	+	0
3 =	0	+	0	+	0	+	0	+	1	+	1
4 =	0	+	0	+	0	+	1	+	0	+	0
5 =	0	+	0	+	0	+	1	+	0	+	1
6 =	0	+	0	+	0	+	1	+	1	+	0
7 =	0	+	0	+	0	+	1	+	1	+	1
8 =	0	+	0	+	1	+	0	+	0	+	0
.					.						
.					.						
61 =	1	+	1	+	1	+	1	+	0	+	1
62 =	1	+	1	+	1	+	1	+	1	+	0
63 =	1	+	1	+	1	+	1	+	1	+	1

Table 1.1 Numbers 0 - 63 (base ten) expressed in binary (base two).

In digital systems, we can organize the transfer of information in two ways: *parallel*, in which the words are sent complete, one word at a time, and *serial*, in which the words are sent one bit at a time. In parallel systems each bit is carried by its own circuit, so that if we are using 16-bit words to represent information, we have 16 circuits. Examples of parallel systems are the information pathways which carry *data* and *addresses* (storage locations for information) inside computers. These are called the *data bus* and *address bus*, respectively. Alternatively, serial systems, such as those that are sometimes used to connect computers together, use single circuits and send the bits one at a time through these circuits. This latter approach leads to simpler systems but is much slower. The speed of transmission of digital information in serial form is called the *baud rate*. There are also systems which combine both of these approaches, as a 32-bit word may be put on a 16-bit bus by cutting it in two and sending 16-bit sections at a time.

When electrical signals represent numbers in binary form, *logical operations* and *arithmetic operations* may be performed on them. Logic involves "if-then" operations (for example, if input A is logic 1 and input B is logic 1, then the output is logic 1), whereas arithmetic is merely the addition, subtraction, multiplication, and division of numbers. The simplest way of performing numerical operations on binary numbers is to use the logic principles of *Boolean algebra*. There are three main operations in Boolean algebra which may be combined to create the common algebraic functions of addition, multiplication, etc. The Boolean operations are OR, AND, and NOT (Table 1.2). Note that the symbols used should not be confused with common algebraic symbols; "+" does not mean "add" in Boolean algebra. AND OR have *complementary operations* called NAND (NOT AND) and NOR (NOT OR). These are the logical *inverse* of AND OR. The electrical circuits that perform these operations on electrical signals are called *gates*. Individual gates may be used to perform logical operations; for instance, an AND gate may be used to detect whether both of two switches are closed as part of a machine's safety system. Combinations of gates are used for more complex logical operations and to perform mathematical operations, as we will see later in this text. We may also store the numbers in electrical form in memory circuits. This function is also discussed later.

OR (+)	AND ($\cdot$)	NOT ($\overline{}$)	NOR ($\overline{+}$)	NAND ($\overline{\cdot}$)
$0+0=0$	$0 \cdot 0 = 0$	$\bar{0} = 1$	$\overline{0+0} = 1$	$\overline{0 \cdot 0} = 1$
$0+1=1$	$0 \cdot 1 = 0$	$\bar{1} = 0$	$\overline{0+1} = 0$	$\overline{0 \cdot 1} = 1$
$1+0=1$	$1 \cdot 0 = 0$		$\overline{1+0} = 0$	$\overline{1 \cdot 0} = 1$
$1+1=1$	$1 \cdot 1 = 1$		$\overline{1+1} = 0$	$\overline{1 \cdot 1} = 0$

Table 1.2 Logical operations used in Boolean algebra.

Applications of ICs

To finish this chapter, it is fitting to discuss some of the applications of ICs in the information age. Unfortunately, it is possible to look at only a few representative applications, as the technology is so pervasive that several books could be written on this subject alone. To make the task easier, we will divide the applications into three categories: analog, digital, and combinations of these forms.

Analog circuits surround us in our everyday lives. Most homes have television and radio receivers and stereo music systems. These are good examples of analog systems containing analog ICs. There are analog ICs in all of these devices which operate at audio frequencies to amplify the speech and music signals. These signals arrive to the system from the antenna (television/radio), tape head, cartridge, etc. (stereo) in a very weak form and must be amplified so that we can hear them through the loudspeaker. In the case of television, many of the ICs have to operate at ultra-high frequencies (UHF) in order to process the UHF carrier which brings the signal, picture, and sound from the transmitter to the antenna. There are ICs in the television receiver which must amplify the tiny incoming voltages and then *demodulate* (separate) the signal from the carrier and sort out the vision component from the sound component. The telephone is also basically an analog system, complete with audio amplifiers and a system of audio tones which are used to set the switches at the exchange in Dual Tone Multi-Frequency (DTMF) schemes. These exchange switches are now becoming digitally controlled, but we will return to the combination of analog and digital systems later. Many common appliances (unless they contain only an electric motor) and electronic systems, whether analog or digital, have a vital analog circuit—the *power supply*. Since most circuits containing transistors operate at around five volts, we cannot connect them directly to the 115 volt electricity supply. In addition, most circuits operate with *direct current* (dc), that is, a steady voltage, whereas the electricity supply is *alternating current* (ac), which changes from positive to negative polarity 60 times per second. The power supply circuit converts the 115 volt ac to five volt dc by first using a *transformer*, an ac component which reduces the voltage, and then a *bridge rectifier*, a circuit which contains diodes to turn the ac into a messy dc. A large capacitor, acting as a charge storage component, is used to smooth the dc and then a *regulator* circuit ensures that the voltage remains constant. This type of circuit illustrates a major drawback with ICs for analog applications: transformers and large capacitors cannot be integrated. Therefore, these have to be discrete components, connected externally to the integrated circuit.

Although commercially available digital systems are relative newcomers compared to analog systems with active components, which basically appeared in the form of radios and intercoms in homes and offices in the late 1920s, they have been so thoroughly integrated into society that many people would find life difficult without them. The first digital systems to gain widespread popular use were calculators, which were originally based on discrete-component technologies but soon became highly portable when suitable ICs became available. These accept numerical *input* from a *keyboard*, process the data in numerical form using digital logic, and *output* the results on an appropriate *display*. The circuitry used for these units combines digital representations of numbers in some predetermined fashion. Sections of *hardware*, permanently "hard"-wired circuitry, perform the particular arithmetic operations they are designed to perform at the press of a button on the keyboard, like addition and multiplication. This type of digital circuit is also known as *asynchronous logic* as it does not require a system clock (see later) to regulate its operation.

Eventually, programmable calculators, in which a user-defined, stored program could be used to initiate the functions in the arithmetic unit in a particular sequence, became available. This meant that memory circuits were also required to hold the program and any data the program needed or produced. A digital circuit also had to be included which could interpret the program *instruction set*, including both arithmetic functions (e.g., "X Y = Z" means "multiply the number in data memory location X by the number in data memory location Y and put the result in data memory location Z") and logical functions ("IF X > Y THEN GOTO M" means "if the number in data memory location X is greater than the number in data memory location Y, then go to position M in the program memory and carry out that instruction"). This generally is how our large *mainframes* and smaller *personal computers* (PCs) work too in that the instructions on what to do with the data are contained in the program held in program memory. Each instruction is carried out sequentially, and the speed performance of computers is thus often quoted in millions of instructions per second (MIPS). Computers are based on *synchronous logic* in that a *system clock* is used to regulate all the operations in the system. The clock provides a train of pulses (analogous to the beat of a metronome), typically 20 million per second in a modern PC, to which all operations are synchronized. This is necessary, as slight delays in information transfer which are inherent in electrical systems could result in an instructions not being fully executed before the next one is invoked. To allow the use of an easy-to-understand computer language, *compilers* are often used in mainframes and PCs to translate from the higher-level language (Fortran, Pascal, C, etc.) to something the computer can understand—*machine code*. In microprocessors, the permanently stored guide on how to respond to instructions is called the *microcode*. This allows the user to use simple *commands* or instructions to initiate fairly complex functions. When functions are not available in the instruction set, they have to be made up, using the available functions by activating an appropriate sequence of commands in the program. This program or *software* approach is inevitably slower than a custom-designed hardware approach, as each instruction has to be fetched, interpreted, and carried out sequentially in the software case, whereas the function may be carried out almost instantaneously in the hardware case. Thus, the software approach is used with flexible systems which may be used for a number of applications, and the hardware approach is used when speed is of the essence. Another task which digital systems are particularly good at is counting; this skill is embodied in digital watches and clocks. A *quartz crystal*, similar to that used in some computer clocks, provides a constant number of pulses per second; these pulses are counted and converted to a time readout by digital ICs.

We are now witnessing the marriage of analog and digital systems for many applications. The reason is that it is frequently easier to control or manage a situation involving analog inputs and outputs using a digital system which can make logical decisions. For example, a central computer could be used to monitor temperature (an analog variable) in a large office complex. The computer hardware and its user-defined software could be set up in such a way that the computer scans and measures the temperature in each area, inputs the values into the control program, and sends a signal to the air conditioning units to raise or lower the temperature in each area accordingly. The flexibility of computer programming could also allow relaxation of temperature control to save energy when the offices were not occupied. The problem with this and related approaches is that temperature is an analog variable and is therefore incompatible with the digital circuitry of the computer. Therefore, if this scheme is to work, the analog signal from the temperature transducers has to be converted to digital form (Fig. 1.4). Conversely, the digital message from the computer to the air conditioners has to be converted to an analog signal in order to operate steam or coolant valves. The circuits which perform these tasks are *analog-to-digital converters (ADCs)* and *digital-to-analog converters (DACs)*, respectively.

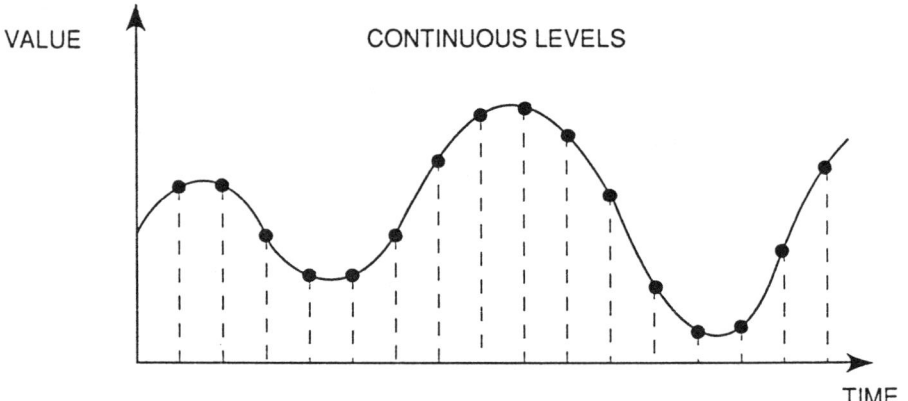

(a) Analog signal showing sampling points

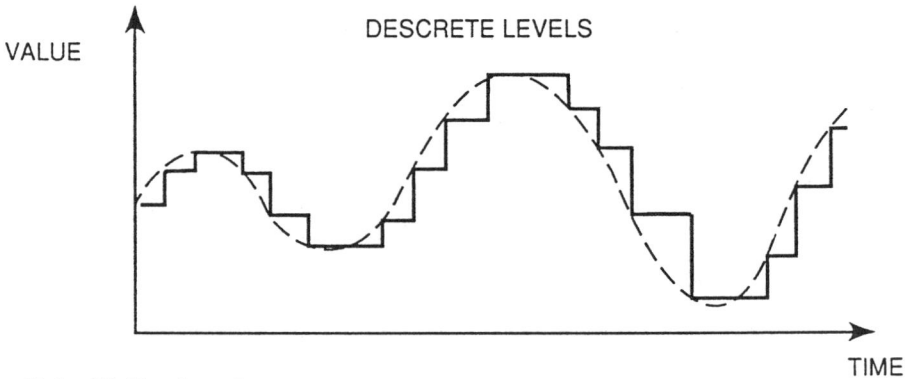

(b) Digitized version

Figure 1.4 Analog to digital conversion.

An ADC *samples* the analog signal and converts the voltage sample it collects to a number which depends on the magnitude of the voltage. The higher the sampling frequency, the more accurate the digital version of the signal becomes. The DAC works the other way, taking a number in binary form and outputing a corresponding voltage. Perhaps some of the most interesting uses of ADC in recent times are the *compact disc* and *digital audio tape*. In these schemes, the music is *digitized* using a very high quality/high sampling rate ADC, and the digital version is put on a magnetic tape. This tape may be used to create tape copies in the case of digital audio tape or to make a digital compact disc, both of which retain the sound information in digital form. During playback, the digital information is read by a tape head in digital audio tape and by a laser in the compact disc and then reconverted back to an analog signal before being amplified and sent to the loudspeaker. The digital approach leads to a much better overall sound quality, as analog copying and playback always adds *noise* to the analog signal, which reduces the quality of the sound. This noise does not appear in digital audio tapes or compact discs as these digital systems basically reject it.

The near future of electronics looks even more exciting. The increased computing power provided by the silicon chip, combined with new developments in computer science, has created possibilities we could only have read about in science fiction books not too long ago. We are now witnessing the use of *artificial intelligence (AI)* in decision and control systems. For instance, programs called *expert systems* can now be used to make decisions based on a *rule base* or *knowledge base* taken from "experts" in the task the computer is required to control, such as the running of a piece of equipment or the diagnosis of a medical ailment.

Some decision-making programs are even using a new type of logic called *fuzzy logic*, which includes the normal "yes" and "no" answers to questions but, in addition, answers like "most of the time" and "hardly ever". It is a lot more difficult to make an electronic machine understand information that is not definitely one thing or another, as computers are made to work digitally (0 or 1), so fuzzy logic (the way you and I think) is a great breakthrough. Finally, totally new architectures are currently being developed which depart from the stored program approach and allow the circuit to develop its own way of dealing with a particular problem through learning by trial and error or heuristics (again, much like you and I). These neural networks are still in the early stages of development but offer the very real possibility of a truly thinking machine.

Summary

The evolution of electronics from the early days of the cumbersome vacuum tube to today's ULSI systems containing millions of components has provided us with the necessary technology for communications and computing in the information age. Analog and digital systems have been totally integrated into society and now surround us in our everyday lives.

2

The Science of
Semiconductors

Conductors, Insulators, and Semiconductors

All materials are composed of atoms. These basically have negatively charged *electrons* which form clouds around *nuclei* composed of positively charged *protons* and uncharged *neutrons*. Unless the atom has been *ionized* (electrons removed or added), there are equal numbers of electrons and protons. In a solid, the nuclei are essentially fixed and cannot move very far; they will vibrate when the material is heated but do not move around except under special circumstances. However, some of the electrons may be held very loosely, depending on the material, and may break free to roam around. Because of this effect, there are essentially three types of materials in the electronic sense: *conductors*, *insulators*, and *semiconductors*.

In conductors, such as most metals, there are many electrons which have been freed from their nuclei at room temperature. These electrons wander around within the material in a random fashion. However, since they are mobile, they are available for *conduction*. This means that if an *electric field* is applied by connecting the conductor to a voltage source such as a battery, the electrons will move through the material in the direction of the field, attracted by the positive voltage or *potential*. Note that an electric field will extend from a region which has a relatively positive voltage to a region which has a relatively negative voltage. The field can exist even when there is nothing between the regions, but current can flow only if electrons are there. The electric field supplies the force to drive the electrons. In the case of our conductor, the flow of electrons constitutes an electric current. (What is also interesting is that if we heat the conductor to a high enough temperature, the electrons gain enough energy from the heat to escape from the surface altogether. This so-called *thermionic emission* is the way vacuum tubes, including CRTs, work.)

In insulators, such as many ceramics or plastics, the electrons are tightly held by the nuclei; since they cannot move, no current flows under the influence of an electric field. However, when a field is applied, the positive and negative charges within the insulator tend to become displaced slightly to create *dipoles*, which are merely regions with a positive charge at one end and a negative charge at the other. This is why we often call these materials *dielectrics*. It is possible to make insulators conduct to some degree by heating them so that the electrons gain enough energy from the heat to break free from their nuclei. Alternatively, a very high electric field (high applied voltage) may drag the electrons from their nuclei. Both of these conditions can result in the destruction of the material and therefore should be avoided; this is generally called *breakdown*.

In simple terms, semiconductors such as silicon fall between conductors and insulators. At very low temperatures, when the electrons have little energy and stay by the nuclei, they are good insulators, whereas at high temperatures there are many free energetic electrons and they are good conductors. At room temperature, a few electrons are free to conduct, so semiconductors are neither good insulators nor good conductors (hence the name). Semiconductors also differ from normal conductors in that there are two types of *charge carriers* (entities which carry a charge through the material) available for conduction, albeit at small quantities for an *intrinsic* (pure) semiconductor at room temperature: *electrons* and *holes*. We will discuss these carriers further later in this chapter.

To examine semiconductors further, we must now look at the *periodic table of the elements*. This is merely a chart of all the elements, the basic building blocks of chemistry, arranged to show the relationship between the materials. Table 2.1 shows part of the

II	III	IV	V	VI
	B	C	N	O
	Al	Si	P	S
Zn	Ga	Ge	As	Se
Cd	In	Sn	Sb	Te
Hg	Tl	Pb	Bi	Po

Table 2.1 Part of the periodic table of the elements, showing semiconductors and dopants.

periodic table. Naturally occurring *elemental semiconductors* (single elements) at room temperature are the *group IV* elements *silicon* (Si) and *germanium* (Ge).These elements have a valency of four which means essentially that the atoms can make four bonds with other atoms. It is possible to combine *group III* and *group V* elements to form *compound semiconductors* such as *gallium arsenide* (Ga + As = GaAs) or *indium phosphide* (In + P = InP). In the electrical sense, these *III-V* (said "three-five") semiconductors behave in much the same way as the group IV semiconductors. It is also possible to create *II-VI* compound semiconductors such as *cadmium telluride* (CdTe) or *ternary* (three-element) compound semiconductors such as *aluminum gallium arsenide* (AlGaAs) or *mercury cadmium telluride* (HgCdTe). These materials are all semiconductors in that a few electrons (and holes) are free for conduction at room temperature, but they possess different properties from each other, such as sensitivity to different wavelengths of light.

Silicon is one of the most abundant elements in the earth's crust, and is relatively easy to refine and process to create devices and circuits. Compound semiconductors such as GaAs or InP do not have these advantages, so why use them? Because they have properties which silicon does not! The *electron mobility*, the speed at which an electron can travel through the material, a factor vital in the speed of operation of a device, is as much as six times higher in GaAs than in

Si at room temperature. Also, many compound semiconductors are able to *emit* or *detect light* under certain conditions and therefore may be used in *displays* and *optical communications*. It is interesting to note that the first commercial discrete diodes and transistors were based on germanium rather than silicon but this element is rarely used in mainstream IC applications today due to problems of processing the material.

In almost all cases, the semiconducting materials used in devices and ICs are *crystalline*. In crystalline materials, the atoms bond together in such a way as to arrange themselves in a highly regular fashion, forming a geometrically perfect *lattice*. Figure 2.1 shows the basic crystal structure of silicon in which each silicon atom bonds *tetrahedrally* (its four bonds spaced equally around the sphere of the atom) to four of its neighbors. Crystal lattices in materials possess interesting electrical properties.

Electrons and Energy

If we were to take one atom in isolation, we would see that the electrons around its nucleus can only possess certain *single* or *discrete energies*. The values of the *energy levels* are determined by the laws of a branch of physics called *quantum mechanics*. Associated with the energy levels are *states* which may be occupied only by one electron at any one time. When many atoms are in close proximity, as in a crystal lattice, the discrete levels split up to form *energy bands*; the available states can have values of energy which fall within these bands but cannot have energies which are outside them. Each band has a limited number of states and therefore can only accommodate that same number of electrons (strictly, one electron per state). The highest energy bands in a material are the *conduction* and *valence bands*. These two bands are separated by a range of forbidden energies called the *bandgap*. The bandgap is an important physical quantity in materials, as it determines how the material behaves electrically when heated or illuminated with light.

In order for an electron to take part in conduction, it must have enough energy to allow it to exist in the conduction band where it is free to move away from its original nucleus around the entire crystal lattice. Electrons in the valence band are more tightly held by the nucleus (they are physically closer to it). It is these valence electrons which take part in the chemical bonds holding the atoms together. In

Silicon ——
Atom

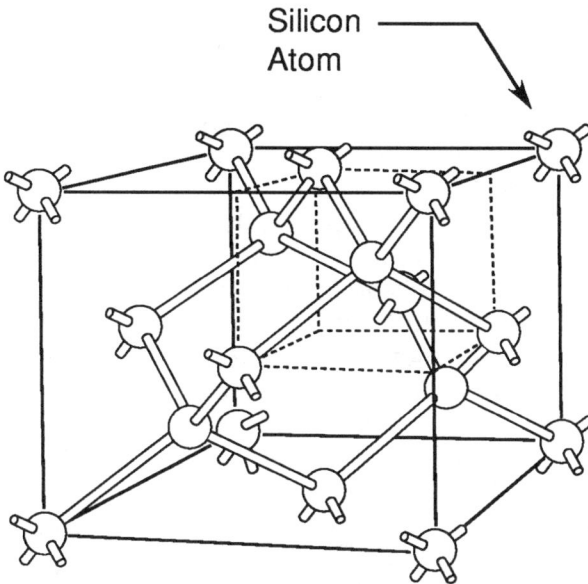

Figure 2.1 The crystal lattice of Si.

addition, there must be free (unoccupied) states in a band to give an electron "room" to move around. Therefore, there can be no electron movement and hence no conduction in a valence band which is completely occupied by electrons.

Figure 2.2 shows the energy bands for a conductor, an insulator, and a semiconductor. Note that in these *energy band diagrams* the vertical axis has the units of energy; the higher the position in the diagram, the higher the energy. In a conductor, the structure of the bands is such that there are many electrons in the conduction band at room temperature. In an insulator, there are no electrons in the conduction band and the valence band is full at room temperature. In addition, the bandgap between the valence and conduction bands effectively prevents electrons from the valence band from getting into the conduction band unless we give them a huge amount of energy by heating the material (remember breakdown). In a semiconductor, however, the bands are closer together, so that even at room temperature, some electrons have enough energy to leave the valence band and reach the conduction band. The number of electrons which are able to do this at a particular temperature is once again given by the laws of quantum mechanics. Put simply, as the temperature rises, more electrons can cross the gap. That is why a semiconductor becomes a better conductor at higher temperatures.

When electrons leave the valence band in a semiconductor, they leave behind *holes* which are effectively the absence of electrons or empty states. This is called *electron-hole generation*. The curious thing about holes is that they can apparently travel in the material under the influence of an electric field. In fact, it is actually the other valence electrons which have begun to move in the band, since it is no longer completely full. When an electric field is applied, the electrons in the conduction band move toward the positive potential, but so do the electrons in the valence band, since they now have free states to move into. The net effect is that the holes appear to move in the opposite direction, as illustrated in Fig. 2.3. This makes it appear as though the holes are moving. Since they are moving in the opposite direction to the electrons, toward the negative potential, they effectively behave as if they are positively charged. The holes therefore contribute to overall conduction in the semiconductor. Holes move more slowly than electrons, though. In silicon, for example, holes are three times slower than electrons in the conduction band, because the valence band electrons are closer to the nucleus and are more tightly held. Electrons in the conduction band can also lose energy and fall into holes in the valence band. This process, known as *recombination*, reduces the number of conduction electrons and holes. However, electrons and holes are also constantly being regenerated with the help of thermal energy, so that at any temperature greater than absolute zero, unless special circumstances occur, there is an equilibrium number of electrons and holes in a pure semiconductor.

We can view the situation in a more physical way. A two-dimensional representation of an intrinsic silicon lattice is shown in Fig. 2.4 (it is easier to draw it this way). The outermost *valence electrons* are taken up in the process of forming *covalent bonds*, a particular type of bond between atoms which involves the sharing of valence electrons. In a perfect covalent bond in silicon, a valence electron (each silicon atom has four) pairs with a valence electron from one of the four nearest neighboring atoms. This is why we say that silicon has a valency of four. Going back to our energy band diagram representation, this situation corresponds to all the electrons

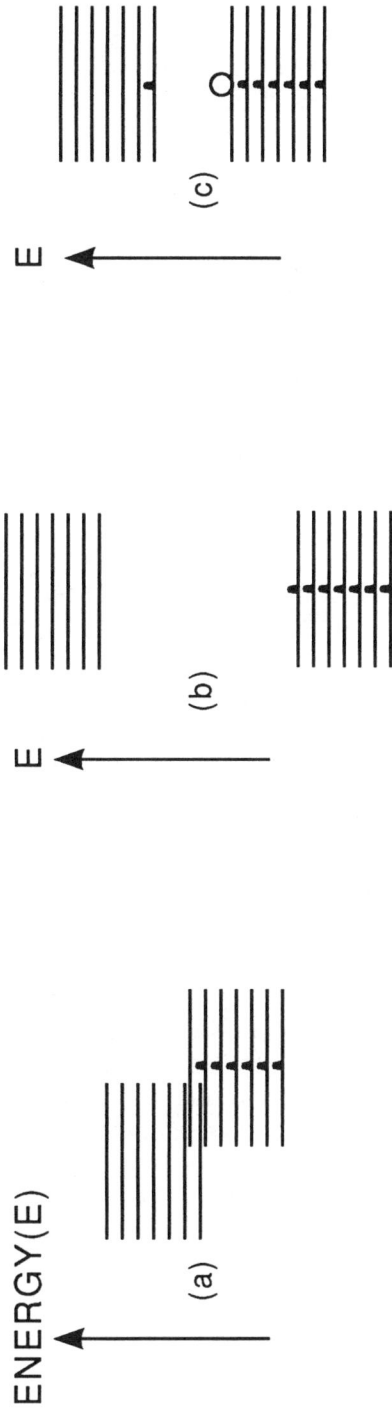

Figure 2.2   Band structures of (a) conductors, (b) insulators, and (c) semiconductors

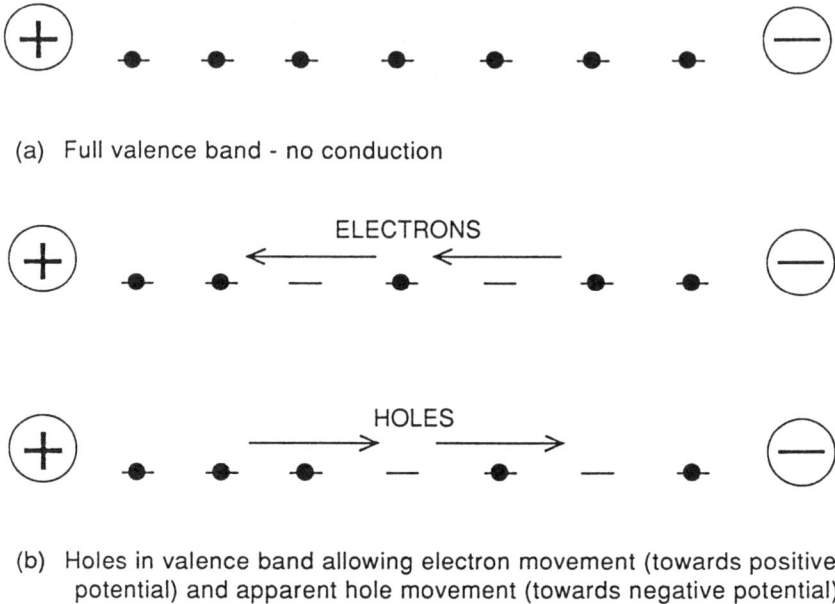

(a) Full valence band - no conduction

(b) Holes in valence band allowing electron movement (towards positive
     potential) and apparent hole movement (towards negative potential).

Figure 2.3 Hole conduction in the valence band.

being in the valence band.   As discussed previously, at room
temperature there is enough thermal energy to allow electrons to
leave the valence band; this results in the situation shown in Fig. 2.5.
The electrons which rise into the conduction band can leave their
nuclei and move elsewhere in the lattice, leaving behind a net positive
charge, a hole, in the process.   (Remember that the atoms start off
with equal numbers of negative and positive charges; therefore, if we
take away an electron, a net positive charge is left.)

To provide more carriers for conduction at room temperature and
thereby change the electrical characteristics, a pure or *intrinsic*
semiconductor may be *doped* to create an *extrinsic* semiconductor.
Doping involves the introduction of certain *impurities* into the crystal
lattice. We will discuss how this is done in Chapter 4. The impurities
or *dopants* are elements which generally have a valency of one more
or one less than the material making up the lattice.   In the case of
silicon, dopants would be group V elements (five valence electrons),
phosphorus (P), *arsenic* (As) or group III elements (three-valence

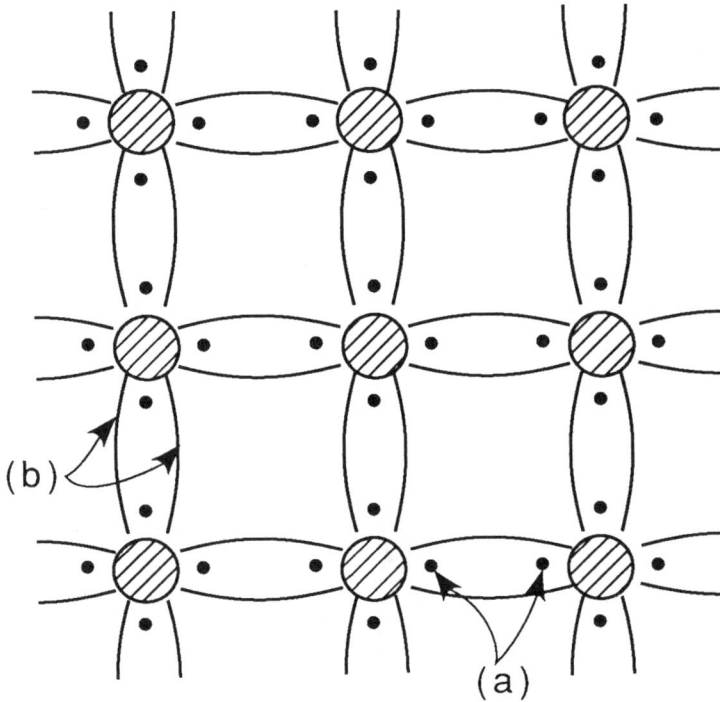

Figure 2.4 Two-dimensional representation of a crystal lattice showing (a) valence electrons and (b) a covalent bond.

electrons), *boron* (B), *aluminum* (Al). For GaAs, dopants may be group II, IV, or VI elements such as *cadmium* (Cd), *tin* (Sn) or *sulfur* (S).

For an impurity to provide extra carriers, it must become *electrically active,* but in order to do this it must occupy a lattice site and bond with the semiconductor. In the case of silicon, phosphorus doping provides extra electrons (Fig. 2.6), as each phosphorus atom, with a valency of five, has one more valence electron than the silicon atom it substitutes for; only four bonds are formed with the four nearest neighboring silicon atoms in the lattice, leaving one unbonded electron. It should be noted, though, that if the extra electron leaves the phosphorus atom by entering the conduction band, it leaves behind a *positively charged, fixed phosphorus ion.* This is not a hole, as it has not arisen from the valence band of the silicon, but it is associated exclusively with the phosphorus ion and cannot move. The extra negative charges from the electrons are thus balanced in the material by the fixed positive phosphorus ions, and the material remains electrically balanced or *neutral*. Phosphorus is therefore known as an *n-type* dopant, as it provides extra negatively charged carriers. Boron, with a valency of three, has one fewer valence electron than silicon; therefore, only three full bonds are made with neighboring silicon atoms when it occupies a lattice site (Fig. 2.7). If the valence electron from the fourth silicon atom joins the boron, the dopant atom becomes a *negatively charged, fixed boron ion*, and a hole is left behind at the silicon (as it has lost a valence electron). Once again, *charge neutrality* occurs as the mobile positive holes are balanced by the fixed negative ions. Boron is thus known as a *p-type*

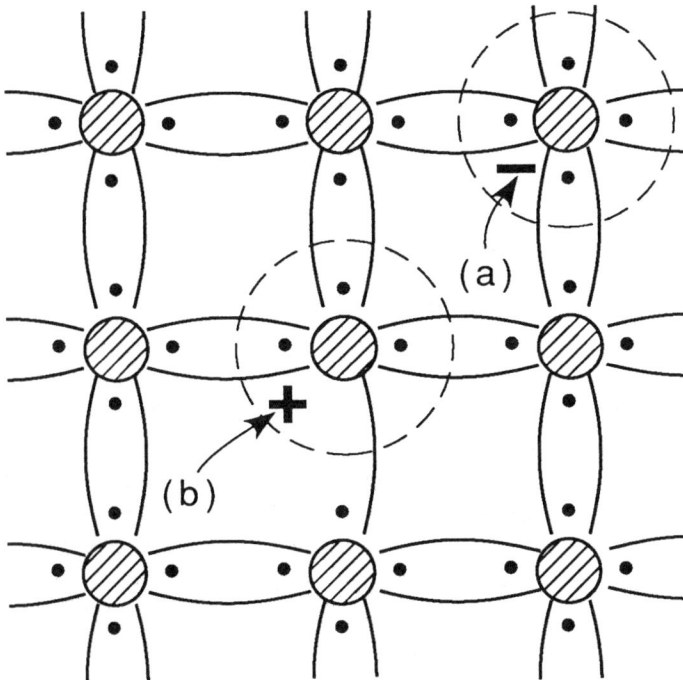

Figure 2.5 Crystal lattice with one bond broken showing (a) a free electron and (b) a hole.

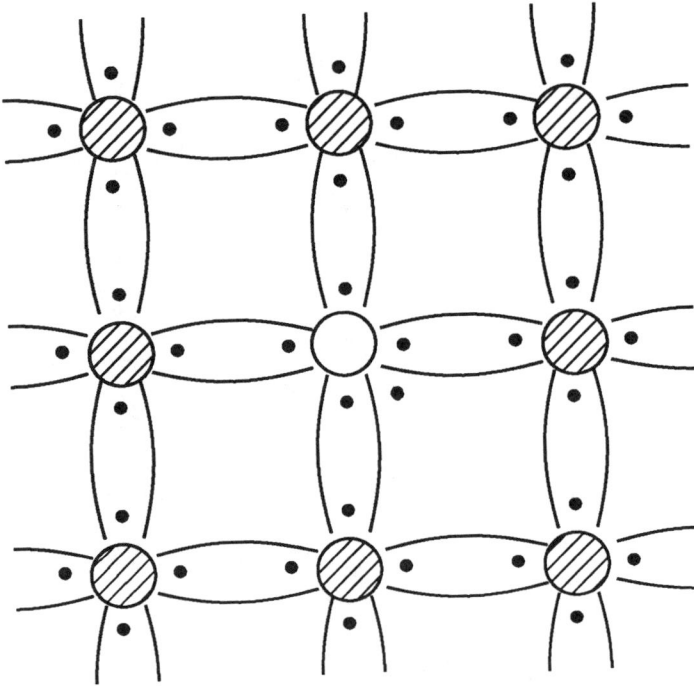

Figure 2.6 Crystal lattice with group V atom.

dopant, as it provides positively charged carriers. In terms of band structure (Fig. 2.8), the impurities provide energy levels within the band gap. This can happen because the dopant atoms do not have the same allowed energy levels associated with them as the semiconductor. In the case of n-type dopants (which are also known as *donors*), the new added states are close to the conduction band and can donate electrons to the band with great ease at room temperature. Similarly, for p-type dopants, also known as *acceptors*, the new gap states are close to the valence band, so that they can accept electrons from the band, leaving behind holes. At typical dopant concentrations, the donor- or acceptor-supplied carriers far outnumber the carriers created by valence band to conduction band transitions; relatively few electrons have enough energy to make this jump, whereas practically all donor or acceptor atoms are ionized at room temperature to create carriers. However, in many cases, we cannot

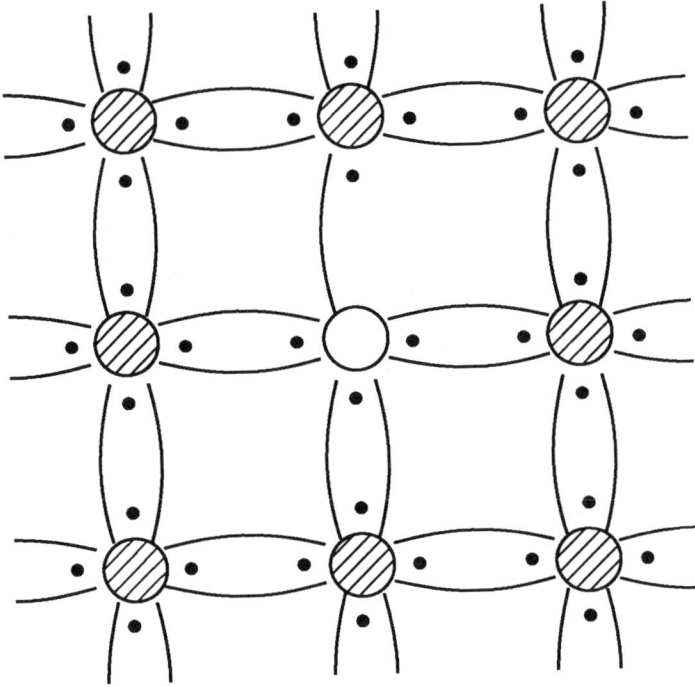

Figure 2.7 Crystal lattice with group III atom.

completely ignore the other electrons or holes which come from the semiconductor rather than the dopant, as many devices are based on the fact that they exist.   It is therefore useful to talk in terms of *majority carriers* and *minority carriers*.   In an n-type material, the electrons are the majority carriers and the holes are the minority carriers.   The opposite is true for a p-type material.

Combining Materials—The p-n Junction

A semiconductor may be doped by a number of means, although as we will see in Chapter 4, *diffusion* or *ion implantation* is usually employed in IC fabrication.   The *bulk* or *substrate* material is typically doped either p- or n-type when the crystal is grown (see Chapter 3),

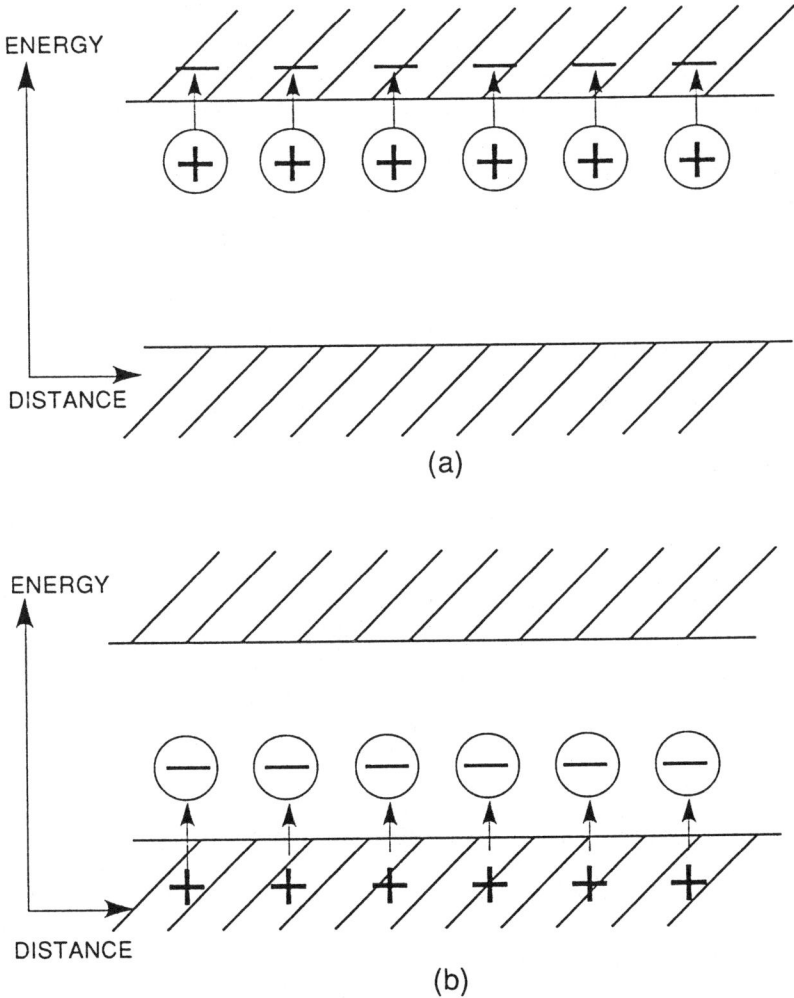

Figure 2.8 Band structures of (a) n-type and (b) p-type semiconductors.

but further impurities of the opposite type may be added to overcome the background doping in order to change the type of the material in selected regions. The concentration of the added material must be considerably greater than that of the background to cancel out or *compensate* for the existing dopant and change the region from n- to p-type or vice versa. Note that in employing heavy doping to create an n-type region, there are, of course, large numbers of electrons in the material from the donor atoms. However, some holes will still be present, although not many compared to the number of electrons.

By introducing a different type of dopant into a substrate in this way, a *junction* of p- and n-type material is formed. This *p-n junction* has a number of interesting properties. Figure 2.9 is a schematic of a p-n junction. When the junction is created, the mobile conduction electrons from the n-type material closest to the junction will cross into the p-type side. They do this because initially there is a high concentration of electrons on the n-type side and a low concentration on the p-type side. This *concentration gradient* allows the electrons to move by *diffusion*, that is, move naturally from high to low concentration. We will return to this concept in Chapter 4. After crossing, they tend to combine with the holes close to the junction. Similarly, the holes from the p-type material cross to cancel out the electrons. This movement of carriers leaves behind the fixed, ionized dopant atoms, creating a charge imbalance on both sides of the junction which causes an internal electric field to be set up (something like an internal battery, but not quite, as by the laws of thermodynamics, we could never draw power from it). This *built-in voltage*, typically around 0.7 volts for silicon, acts to move electrons away from the p-type side and effectively cancels any further diffusion of charge; an *equilibrium* is reached. So, it is only the carriers close to the junction which are affected by the recombination. Since the electrons and holes which have moved across the junction have combined, they are no longer available for conduction. The junction region is *depleted* of carriers, forming a *depletion region* near the

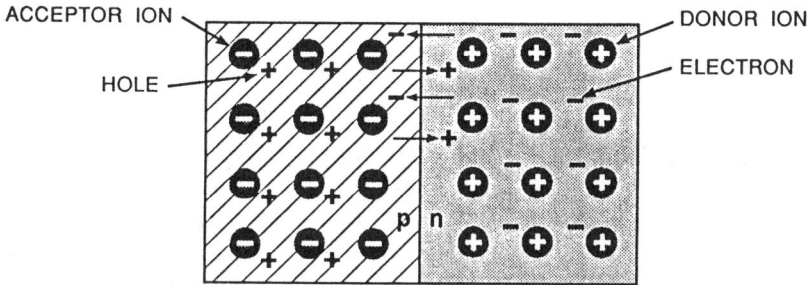

Figure 2.9 Mobile and fixed charges at a p-n junction.

junction. Since most of the carriers near the junction have been taken up by recombination, the depletion region is effectively a *barrier* to conduction. This barrier, whose magnitude depends directly on the size of the built-in potential, is better visualized in the energy band diagram of the p-n junction, which is shown in Fig. 2.10. As may be seen, the act of creating a p-n junction makes the bands of the two types of semiconductor align in an interesting way. There is a definite "hill" where they join, the *potential* (energy) *barrier*. Such potential barriers only allow electrons which have sufficient energy to cross from right to left.

If a positive voltage (the positive terminal on a battery) is applied to the p-type material and a negative voltage (the negative terminal) to the n-type material, the energy band diagram becomes modified, as shown in Fig. 2.11a. The height of the barrier is reduced and more electrons from the n-type material have sufficient energy to cross, moved through the material by the electric field. This condition is known as *forward bias* and a progressively larger current will flow as the applied voltage is increased. Note that the electrons which leave the n-type material are replaced by electrons which go into the material from the contact at the end of the n-type region; in practical terms, this contact is a wire which connects the negative side of the battery or other electrical power source to the p-n junction. The excess electrons in the p-type side are collected by the contact at the end of the region, where they are returned to the battery to complete the circuit. If, however, a negative voltage is applied to the p-type material and a positive voltage to the n-type (Fig. 2.11b), electrons and holes will be attracted away from the junction and the depletion region will become wider. Under this condition, the barrier gets larger with increasing applied voltage, and very few electrons will have sufficient energy to cross. Thus, practically no conduction can take place in this *reverse bias* condition. This *rectifying* characteristic of the p-n junction is extremely important, as it is the basis for many electronic devices. A simple junction such as the one discussed in this section is a semiconductor *diode*. The circuit symbol for this device is shown in Fig. 2.12. Besides being a rectifier, it has a nonlinear relationship between voltage and current under forward bias. This property is useful for *mixing* and *detection* functions in communications.

A further point to keep in mind is that a large number of more complex devices rely on the properties of the p-n junction to function. The next section describes how the most significant of these devices, the transistor, operates.

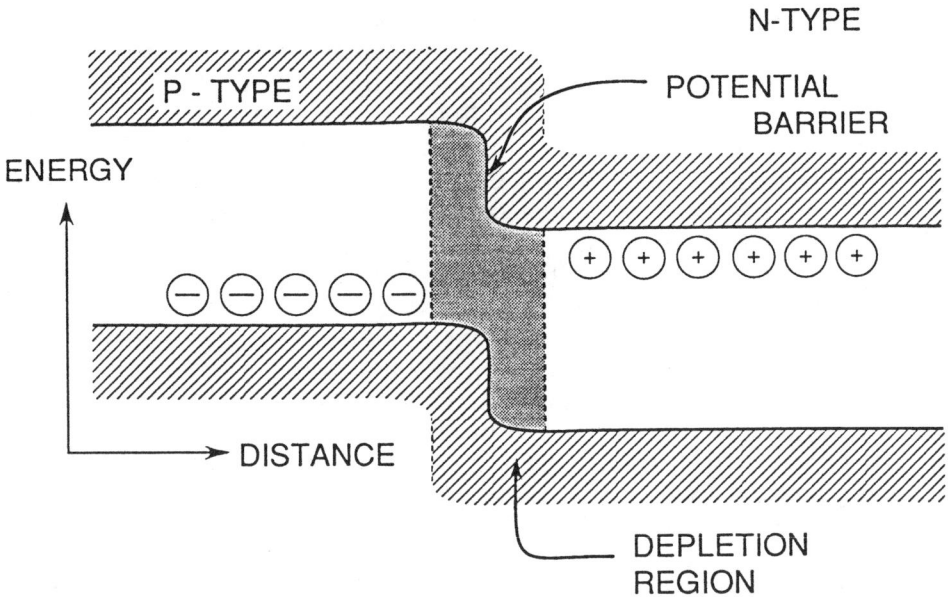

Figure 2.10 Energy diagram of the p-n junction showing the potential barrier.

More Complex Semiconductor Devices

As we saw in the case of the vacuum tube triode in the last chapter, a small voltage can be used to control a large current. A device which uses a small amount of work, energy, voltage, etc. to control a large quantity typically has to have an input for the control parameter and some means of moving the controlled quantity through the device in order to change or *modulate* it. There are many of examples of such devices in everyday life, such as a large valve on a high-pressure water pipe which may be turned by hand (the control input) to regulate the flow of thousands of gallons of water (the controlled quantity). Many of these elements thus fall into the category of *three-terminal devices*, where one terminal acts as a controlling input and the other two take the controlled quantity through the device.

(a) Forward bias case - the barrier gets smaller

(b) Reverse bias - the barrier gets bigger

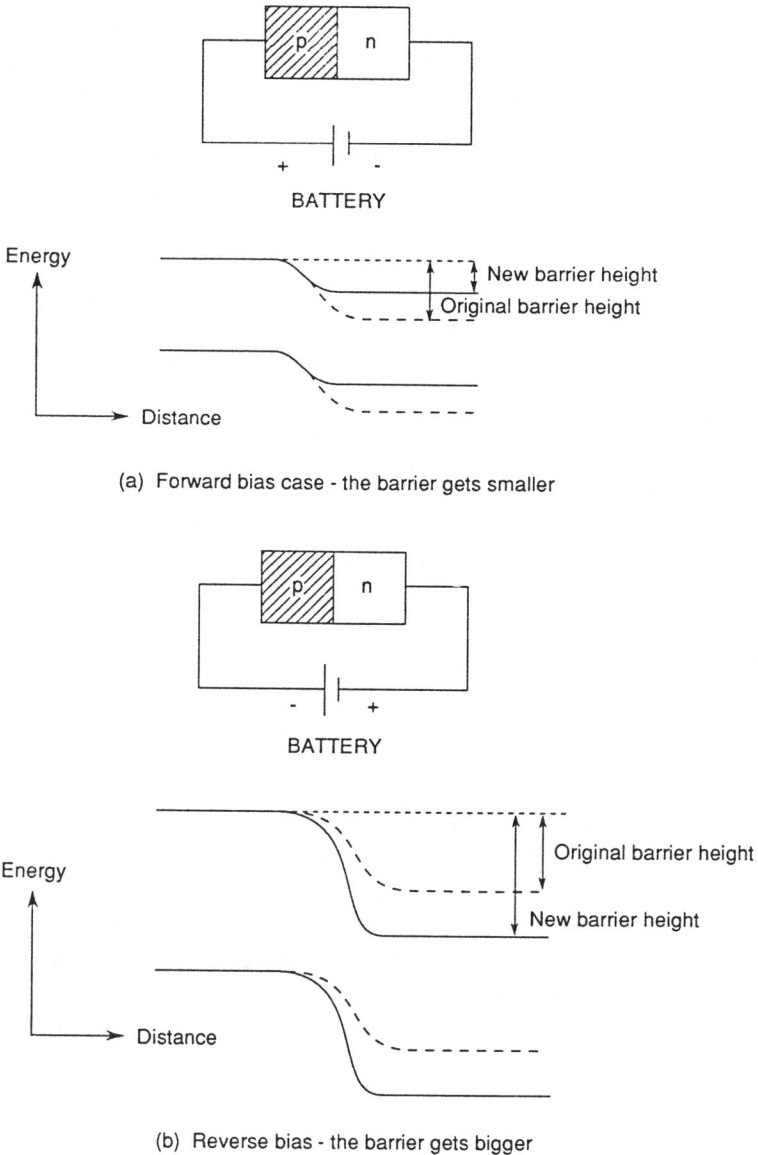

Figure 2.11 Energy diagram of a p-n junction. (a) Forward bias: the barrier gets smaller. (b) Reverse bias: the barrier gets larger.

Electrically, a three-terminal device may act as an amplifier. For instance, in a particular device, a 1-milliamp (one-thousandth of an amp) change in an input current may create a 50-milliamp change in the output (controlled) current. In this case, the *gain* of the amplifier is 50. As discussed previously, amplification is an analog technique, but three-terminal devices can also act as switches; a change in input condition can turn the device most of the way on or off, allowing digital applications. The most important electrical three-terminal device is the *transistor*. The name arose from the term transfer resistor, as the input current or voltage effectively controls the electrical resistance to current flow in the output circuit. Many different types of transistor in use today; we will examine them here.

The first type we will look at is the *bipolar transistor*—a good place to start, as this was the first type to appear commercially. Schematic diagrams of typical bipolar transistor structures are shown in Fig. 2.13, along with their circuit symbols. As we can see, the device has three regions within it: the *collector*, the *base*, and the *emitter*. The centrally located base is electrically connected to the input circuit, while the collector and emitter are connected directly to the output circuit. When we have an n-type emitter, a p-type base, and an n-type collector, the device is called an *n-p-n* transistor. The other variant is the *p-n-p* transistor, in which the base is n-type and the other two regions are p-type. In either case, the transistor looks like two back-to-back p-n junctions and hence is frequently called the *bipolar junction transistor* or BJT.

If we take our n-p-n transistor and apply a positive voltage to the collector contact and a negative voltage to the emitter contact, with no voltage on the base, no current will flow through the device from collector to emitter. This is because, under these conditions, the *collector-base* junction has a positive voltage on the n-type side. Therefore, it will be reverse biased and, like the reverse-biased diode discussed in the last section, will not allow current to flow. However, if we now apply a positive (with respect to the emitter) voltage to the

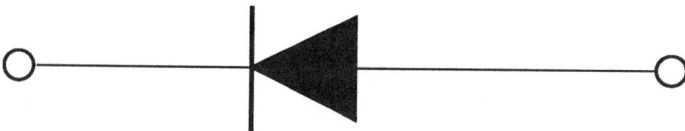

Figure 2.12 Circuit symbol for a diode.

base, we ensure that the base-emitter junction is forward biased and that current flows from the emitter region to the base. Remember, this current is merely a flow of electrons from the n-type material supplied by the (negative) emitter contact, which crosses the base-emitter junction under the influence of the electric field. Now, if the base region is made very thin (much less than 1 mm), the electrons which have been *injected* into the base by the forward biasing will reach the reverse-biased collector-base junction, but they are now at the top of the collector-base barrier, so they will be swept into the emitter region. It is important to realize that the electrons which enter the base from the emitter are compelled to cross the collector-base junction because of the large potential difference of this reverse-biased junction; they are not trying to climb the barrier, as in the case of the reverse-biased diode we examined earlier, but they are already at the top, ready to roll down the potential hill. The more electrons injected into the base from the emitter, the more are available to cross the collector-base junction. Not all electrons make it through the device; a few are lost to recombination with holes in the base region. The loss of these holes is made up by holes which enter the base region from the base contact. This means that a small current has to flow in the input circuit. Once the electrons are swept into the collector, they may leave the device via the collector contact.

So we can see that, with no voltage (or a negative voltage, by the same reasoning used above) applied to the base, no current flows through the device. With a positive voltage on the base, current flows. The higher the base voltage, the more current crosses the forward-biased base-emitter junction and the more current ultimately flows through the device. We have achieved control of an output parameter by an input parameter. If we were to consider a p-n-p device, the above approach would hold, except that we would have to reverse all the polarities (positive becomes negative) and discuss hole movement or *transport* rather than electron transport and vice versa. Incidentally, the device is known as a *bipolar transistor*, as both electrons and holes are involved in its operation; in the n-p-n case, electrons flow in the collector-emitter circuit, while holes flow into the base. The relationships between voltages and currents in BJTs will be discussed in Chapter 5.

The second type of transistor we will examine is the *junction field effect transistor* or JFET. The schematic diagram of a typical JFET is shown in Fig. 2.14. In this case, we have a block of n-type material which has a p-type region at its surface. The input is connected to the p-type region by a large contact to form what is called the *gate*, analogous to the base of the bipolar transistor. Contacts at either end of the n-type region form the

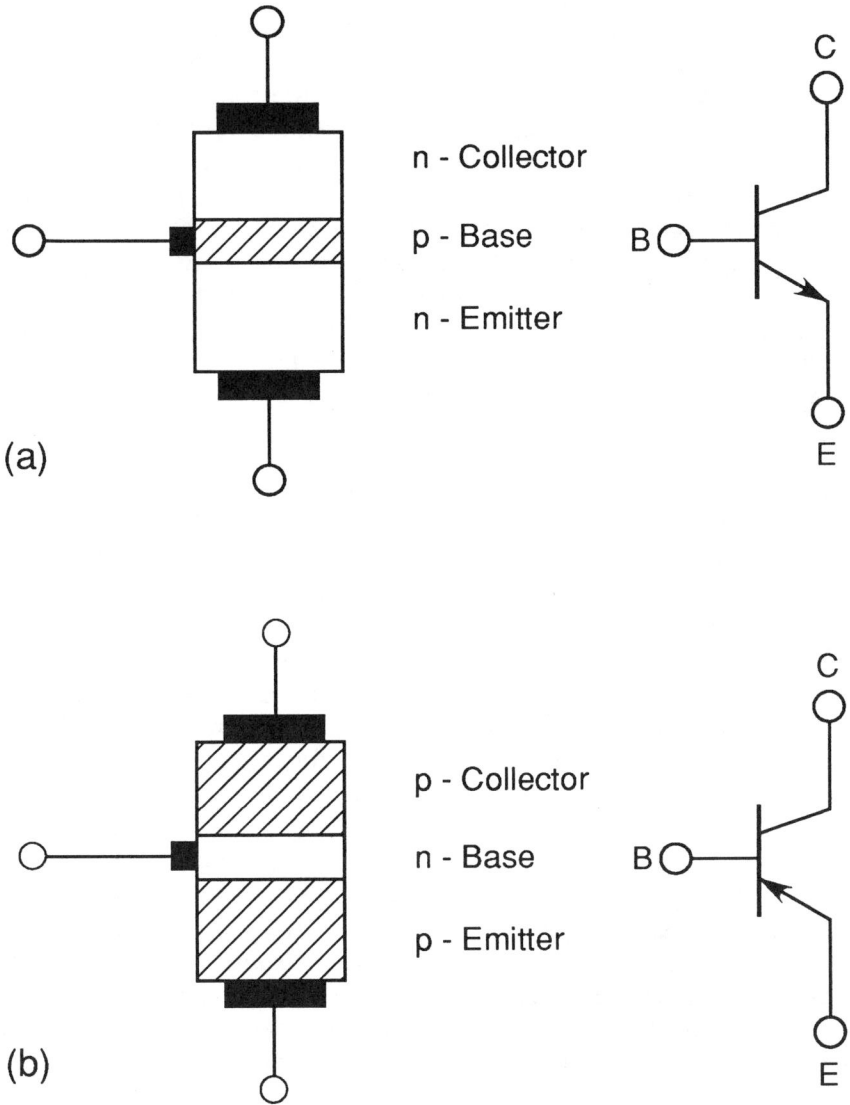

Figure 2.13 Schematic and circuit diagrams for (a) an n-p-n and (b) p-n-p transistor.

source and *drain* areas. This device merely relies on the properties of the reverse-biased junction to operate. With zero voltage applied to the gate, current flows form a drain to the source, as there is no p-n junction between these contacts, only the n-type material. With a negative voltage applied to the gate, the p-n junction at the surface is reverse biased and a depletion region exists there. Note that very little current flows into the gate in the reverse bias condition, typically less than the input current for a bipolar device. The depletion region is actually maintained by the *electric field* created by the voltage on the gate, hence the name of the device. We encountered the concept of the depletion region when we discussed the p-n junction, but we have much more to learn about this phenomenon.

A critical factor we must be aware of is that, due to the smaller number of charges per unit volume, a lightly doped semiconductor will form a wider depletion region than a heavily doped material for the same applied voltage. Also, as the applied voltage increases, the width of the depletion region increases. Therefore, if the n-type part of our JFET is lightly doped, the depletion region can extend far into this area. In addition, if we make the drain positive, the source must be made more negative to induce current flow through the n-type region. Therefore, the depletion region will be much wider at the drain end because of the larger voltage difference there. The resulting shape of the depletion region is also illustrated in Fig. 2.14. Since a depletion region is effectively a zone devoid of (majority) carriers, the

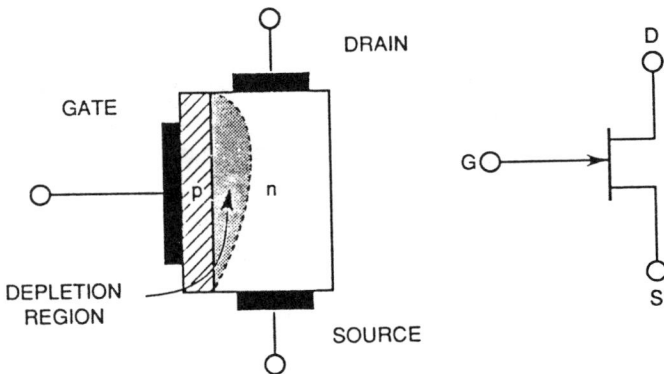

Figure 2.14 Schematic and circuit symbol for a JFET.

electrical resistance is very high. As the gate voltage increases in the negative direction, the depletion region extends further into the n-type material.   The conducting channel in the n-type material therefore becomes constricted, which in turn reduces the current flow from source to drain. Ultimately, the channel becomes *pinched off* and very little current flows.

Since the only carriers supporting current flow in this case are electrons (not both electrons and holes), the device is termed *unipolar*. A device of this type could also be created from an n-type junction into p-type material, but we would have to reverse the polarities. One critical difference between the JFET and the bipolar transistor is that current flows with no applied voltage and is reduced by an increasing gate voltage; put in switching terms, it is *normally on* rather than *normally off*, as in the bipolar case. This and other factors, which we will discuss later, makes this device type a less obvious choice for ICs.   However, in its alternative form, the *metal-semiconductor field effect transistor (MESFET)*, it is widely used in other applications.

The third device type, like the bipolar transistor, is widely used in   ICs.   It   is   another   field   effect   device   called   the *metal-oxide-semiconductor (or silicon) field effect transistor* or MOSFET. This device differs from the JFET in that the metal gate electrode is separated from the semiconductor by a layer of insulator, usually an oxide.   This is where the device gets its name and why it is also sometimes called the *insulated gate field effect transistor (IGFET) or metal-insulator-semiconductor field effect transistor (MISFET)*.   The schematic diagram of a typical MOSFET structure is given in Fig. 2.15. At first glance, this schematic may look somewhat like a bipolar transistor in that the n-type source and drain regions are separated by a p-type region in the middle.   As in the bipolar transistor, if we apply a positive voltage to the drain and a negative voltage to the source, with zero applied voltage to the gate, no current flows from drain to source, as the drain junction is reversed biased. If we apply a positive voltage to the gate, no current flows into the gate, as the gate electrode is insulated from the semiconductor.   This device therefore takes the least input current of the three we have examined.

Since the gate electrode has a voltage on it, an electric field will extend through the insulator to affect the underlying semiconductor. The field can actually change the nature of the p-type material at the surface where it meets the insulator.   As the positive voltage increases, the positively charged holes are pushed away from the surface until a depletion region is formed. If we continue to increase the gate voltage, we will reach a point where the small number of electrons outnumber the holes. This is known as *inversion*, as it leaves the surface looking like an n-type material. Since the inversion layer appears to be n-type, we no longer have a reverse-biased junction at the drain end, and the layer forms a channel linking the drain to the source along the surface. The larger the gate voltage, the wider the inversion layer and the more current can flow through the device. Since the channel resistance will be high, as conduction occurs effectively by the small number of minority carriers, the distance between source and drain must be kept small, typically much less than 0.1 mm, so that the on resistance of the device is as low as possible. The particular MOSFET shown in Fig. 2.15 is called an *n-channel* or NMOS transistor, as the inversion layer appears to be n-type. The alternative is the *p-channel* or PMOS transistor, which has p-type source and drain areas separated by an n-type region. In this case, the gate voltage is negative and the inverted channel appears to be p-type. MOSFETs are also unipolar devices, as only one carrier type is involved in conduction. However, like the BJT, these devices are off with no voltage on the gate and turn on when the gate voltage is increased. (There is a MOSFET variant which is normally on, but we will discuss that in Chapter 6.)

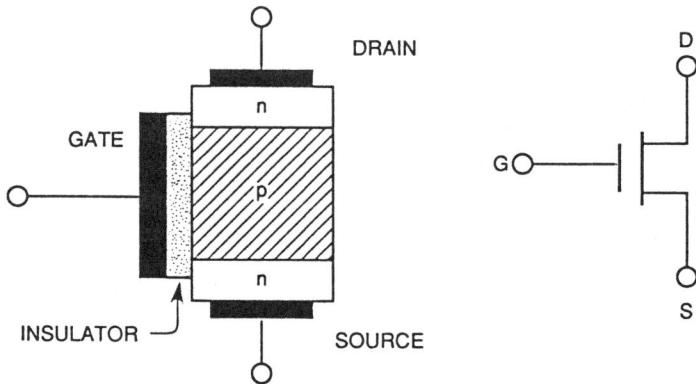

Figure 2.15 Schematic and circuit symbol for an MISFET.

Summary

We have discovered that the fundamental differences between conductors, insulators, and semiconductors in an electrical sense depend on the number of carriers available for conduction. Conductors have many electrons available for conduction, whereas insulators have their electrons tied up. Semiconductors (and some unusual metals) have two carrier types: electrons and holes. Elemental and compound semiconductors exhibit a particular energy band structure in which the conduction and valence bands are separated by a forbidden gap, across which electrons have to have enough energy (from heat) to jump in order to take part in conduction. The electrical properties of semiconductors may be altered using doping; this technique is used to create semiconductor devices. The basic device types are the diode, which allows current flow in only one direction, and the transistor, which uses a small input current or voltage to control a large output current. The two most significant transistor types for IC use are the BJT and the MOSFET.

3

Integration— Bringing the Components Together

Discrete Components

As noted before, devices are the parts which make up a circuit or system. The distinguishing feature of a device is that its behavior may be described in terms of how the currents through it vary with the magnitude of the applied voltages at its terminals. Discrete devices are individual components which may be brought together and interconnected by wires (conductors) to form an electrical circuit to perform a predetermined function which goes beyond the capabilities of each device in isolation. In a circuit made from discrete devices, each device may be taken out and replaced if desired.

In Chapter 2, we discussed: four semiconductor devices, diodes, bipolar junction transistors, junction field effect transistors, and metal-oxide-semiconductor field effect transistors. In their original form, these devices were discrete, existing in their own little boxes or *packages*. They may still be obtained in this form when the application requires it. Transistors are often called *active devices*, as they can amplify electrical signals. Most electrical circuits require a combination of active and *passive* components. Passive components do not amplify as such and usually play a supporting role to the

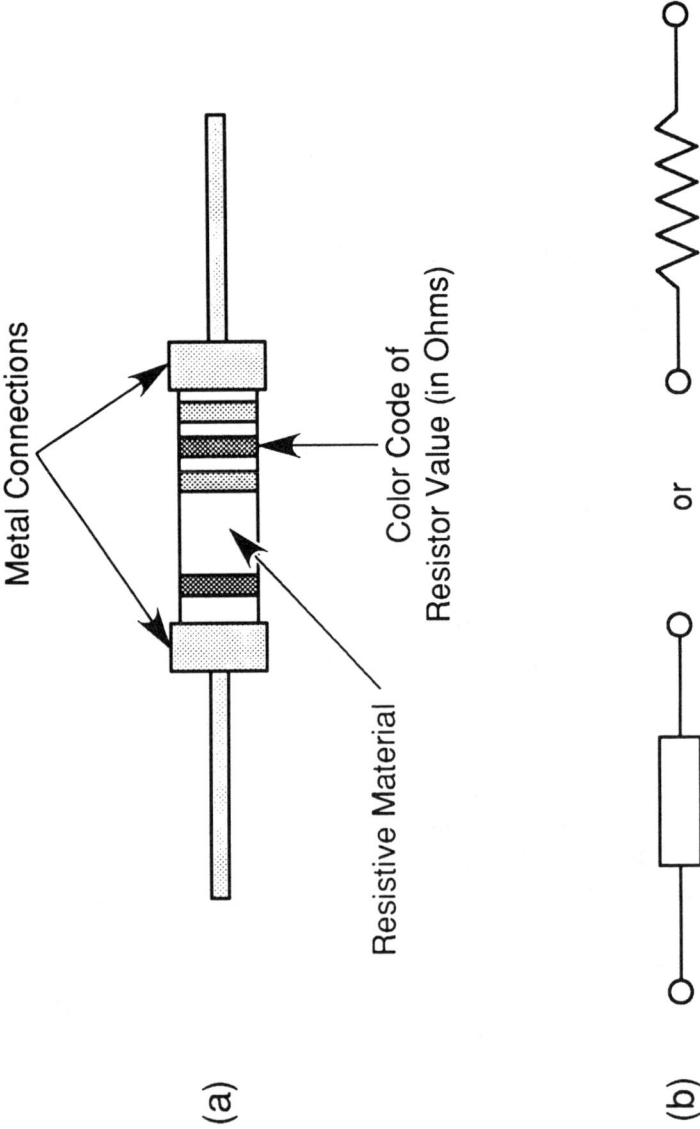

Figure 3.1 The resistor. (a) Diagram of a typical discrete resistor. (b) Circuit symbol used for resistors.

transistors.   The main passive components are resistors, capacitors, and inductors.

Discrete passive components are usually not based on semiconductors.   Resistors (Fig. 3.1) are made of materials which are conductors, but these particular resistive substances, such as carbon, do not have large numbers of conduction electrons.   They are therefore poor conductors and may be used to reduce the flow of current in a circuit.   By *Ohm's law*, the voltage which appears between the terminals of the resistor is proportional to both the current flow through the component and the value (in *ohms*) of the resistor.   This means that resistors may also be used as *voltage dividers* in circuits to split a voltage into two or more values.   A typical analog circuit involving resistors and BJTs is shown in Fig. 3.2.   This is a simple amplifier in which the operating voltages for the bases of the transistors are set by the four-resistor voltage divider which splits the voltage supplied by the power source.   Figure 3.3 shows a typical two-input NAND gate (this term is discussed in Chapter 8) involving MOSFETs and a resistor.   This is a good example of a digital circuit in which the voltage from the power supply is divided by the resistor and the transistors.   We will return to the operational characteristics of these circuits later in this text.

Capacitors are merely metal sheets separated by a layer of dielectric (insulating) material, as illustrated in Fig. 3.4.   Since there is no conducting connection between the *plates*, these elements will not allow current to flow.   However, they can store charge (electrons) on the metal plates.   This ability makes them useful for smoothing out voltage ripples in power supplies, as well as in oscillator and tuning circuits.   The latter type of analog circuit requires a capacitor and a resistor connected as shown in Fig. 3.5.   This section of a so-called tuned circuit will allow maximum current to flow if the voltage across it is varied at a particular frequency determined by the values of the resistor (in units of ohms) and the capacitor (in farads).   It is this property which makes it useful in tuning circuits in radios.   Another related property of capacitors is that they will block steady, invariant voltages and currents, that is, dc, but will allow ac to pass.   We may therefore use them to allow an audio or other ac signal (including rapidly varying digital signals) to pass through a circuit while blocking an unwanted dc voltage.   Charge storage capability is also vital in digital memory circuits, which are discussed later in this book.

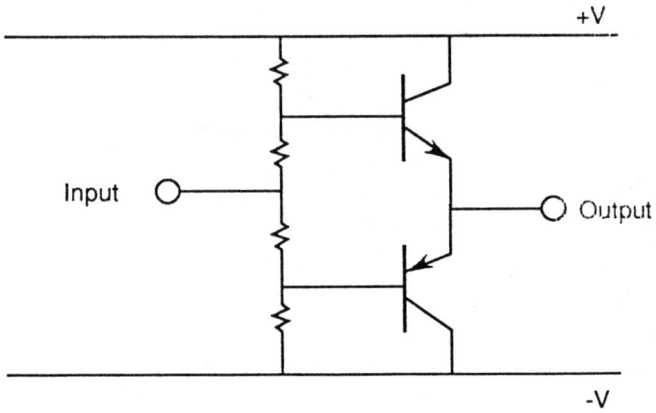

Figure 3.2 Example of an analog circuit—an audio amplifier.

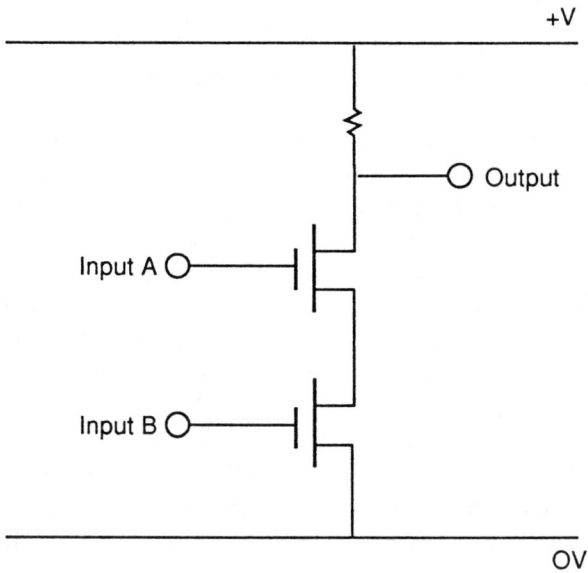

Figure 3.3 Example of a digital circuit— NAND gate.

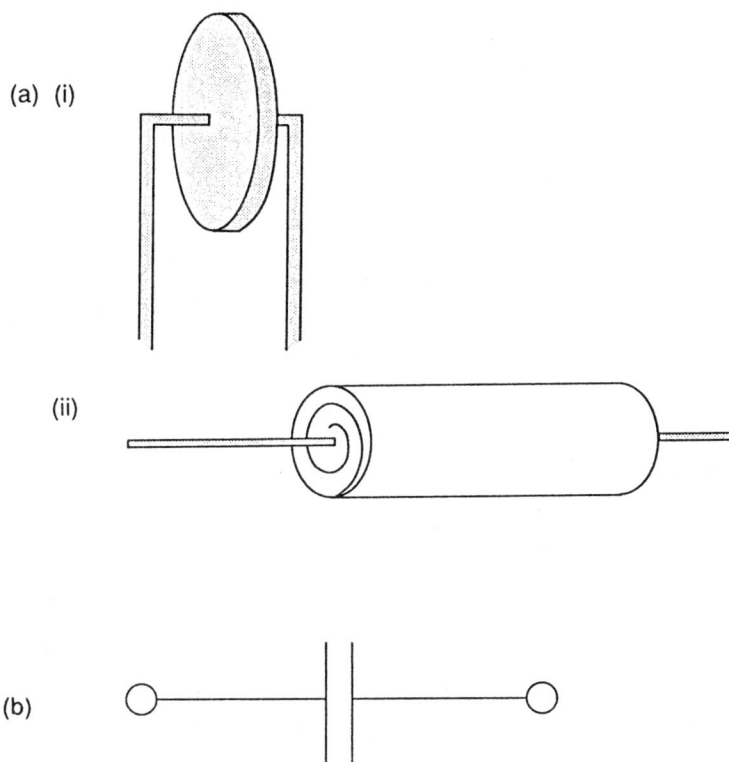

Figure 3.4 The capacitor. (a) Two forms of discrete capacitors: (i) parallel plate; (ii) spiral wound. (b) Circuit symbol used for capacitors.

The third type of passive component is the inductor (Fig. 3.6). It may take the form of a *coil* or *choke* or may be part of a *transformer* was discussed in Chapter 1. Coils are also used in tuning circuits and to generate *magnetic fields*. The transformer, which may be used to convert a low ac voltage to a high one, or vice versa, relies on two coils connected by a magnetic field. This is not amplification, as the *power*, the product of the voltage and current, is the same on both sides of the transformer (or slightly less on the output side due to losses). Once again, dc currents cannot cross the transformer. Although coils find applications in analog communications circuits, generally they are not as common as resistors and capacitors. Further, as mentioned previously, unlike resistors and small-value capacitors, they cannot be easily incorporated in an IC.

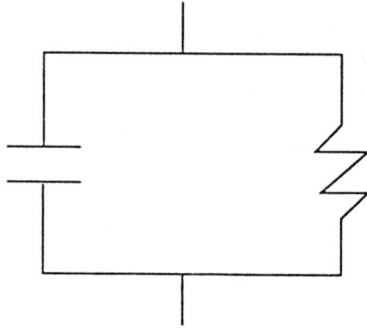

Figure 3.5 Tuned circuit consisting of a resistor and a capacitor.

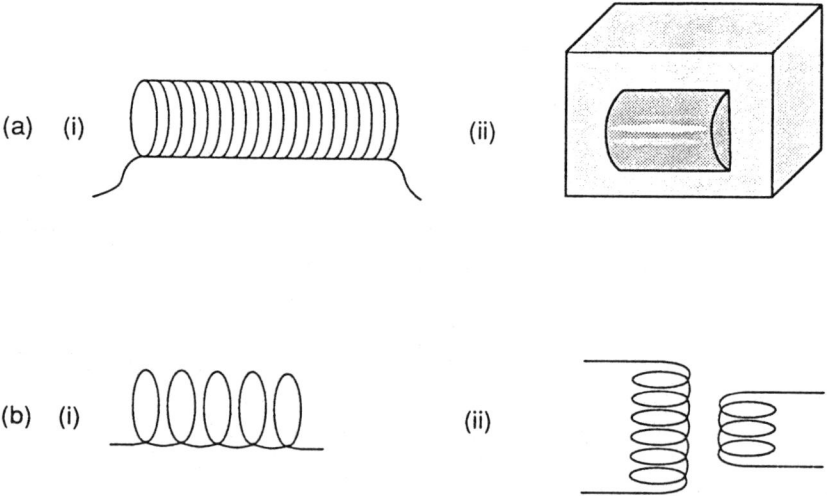

Figure 3.6 The inductor. (a) (i) Diagram of a coil or choke; (ii) a transformer. (b) Electrical symbol for the (i) coil and (ii) transformer.

Printed Circuit Boards to ICs

When we take discrete components and make them into a circuit, we have to join them with conducting materials so that they can interact to perform the function for which the circuit was designed; the wiring carries the electrical currents around the circuit. In practical electronic systems, *printed circuit boards (PCBs)* or *printed wiring boards (PWBs)* are used for this purpose and also to provide mechanical support for the components. Of course, we could join all the components using heavy *copper* wire (a good conductor) and *solder* (a conducting, low-melting-point alloy which may be heated and subsequently cooled to make a semipermanent electrical connection), forming one connection at a time. However, one great advantage of the PWB is that all the connections may be formed simultaneously.

PWBs start life as boards of insulating material, usually fiberglass or something similar, which are covered with a thin layer of copper. The copper is patterned and etched to provide the desired layout of flat wires or tracks to interconnect components. The components, which are manufactured with short pieces of connecting wire or leads attached, are soldered directly to the copper. The patterning is done by *lithography* and *etching*. In many respects, lithography is similar to the process of developing photographic prints in a darkroom. The main steps in the patterning process are as follows:

1. The copper layer is first coated with a *photosensitive* material. This material dries to leave a plastic-like film on the copper.
2. The photosensitive material becomes soluble when exposed to *ultraviolet light*. Therefore, we can transfer an image of the final wiring pattern to this material by the use of a *mask* which has an opaque pattern on it corresponding to the area where we wish the copper wires to be left on the board.
3. After exposure to ultraviolet light, the board is *developed* and *etched* in ferric chloride solution. The copper which is covered by unexposed material remains but the uncovered copper dissolves, leaving the desired wiring pattern in the copper on the board.
4. The remaining photosensitive material is then removed by an appropriate solvent, and the board is drilled to accept the leads of the discrete devices.
5. The components (resistors, capacitors, transistors, etc.) are inserted and soldered in place. The solder holds the components and

creates the electrical connection between the component and the copper track.

The PWB has been used since the days of vacuum tubes. It still is used today, but to connect ICs rather than discrete components. The main drawbacks of the PWB approach are as follows:

- For a complex circuit with many components the board can become unacceptably large. Splitting the circuit into many smaller boards involves great expense in interboard connections. These board-to-board connection systems tend to be expensive, as they have to be assembled, in many cases manually, and high-reliability *connectors* have to be used.
- Reliability decreases drastically with circuit complexity, as the more soldered joints or interboard connections there are, the greater the likelihood of a bad connection, which will affect circuit performance. These bad connections can manifest themselves as weak or worn mechanical contacts in connectors or *dry joints* in solder connections (usually caused by the oxidation of the materials).
- The mass production of boards is labor intensive and requires expensive, complex equipment for component insertion, etc.

To reduce the problems with PWBs, it is possible to fabricate and interconnect most or all of the devices on a semiconductor substrate. The devices are no longer discrete but are now part of an IC.

How do we integrate components in an IC? The answer to this question is not simple; otherwise, this book would be considerably shorter. The secret lies in *planar technology*. It is this which allows the fabrication of complex electrical circuits on a semiconductor *substrate*; the semiconductor becomes the circuit board. As has been mentioned before, with the exception of large-capacitance or high-inductance components, all electronic circuit elements can be fabricated in semiconductor form. In planar technology, all devices are formed by performing operations or *processes* to build up layers of materials with different electrical properties on one flat surface of a semiconductor substrate; hence the use of the word planar. One key advantage is that not only are all the interconnections built up at the same time, as in the case of the PWB, but all similar devices are also built up simultaneously.

There are some other advantages to integration. Since every device does not have to be individually packaged and interconnected on a clumsy PWB, circuits are much smaller and considerably more reliable, as the number of soldered joints and connectors are greatly reduced. Another factor is that since the components in an IC are smaller and closer together, the resistance and capacitance associated with devices and interconnects are reduced, decreasing the time delay for electrical signals. This may not be obvious until one realizes that the combination of unwanted resistances and capacitances in circuits creates a charge storage effect which serves to slow down or *delay* changes in electrical signals. ICs can therefore generally operate faster than their discrete counterparts. Finally, the circuit fabrication technique lends itself well to mass production. Hundreds or even thousands of circuits may be fabricated simultaneously almost as easy as creating a single circuit; therefore, the cost is reduced.

The Semiconductor Circuit Board

The silicon substrates on which the circuits are built are flat, *single-crystal wafers* or *slices* cut from a larger cylindrical crystal. The term single-crystal means that the silicon of the wafers forms a virtually perfect lattice, with few imperfections or *defects* which would adversely affect device performance. The wafer diameter may be ½ inch to 12 inches (300 mm—the latest technology), although 150 mm (6 inches) is common in U.S. production plants (200 mm production quantities are just around the corner). They are called *wafers,* as they are generally only a fraction of a millimeter thick. The largest compound semiconductor wafers (GaAs, etc.) are 3 inches in diameter, but 4 inch wafers will soon become available. Circuits are fabricated on one side of the wafer only. Since a typical IC is less than 1 cm^2 in area, there is room for hundreds of ICs to be formed on a single wafer. Wafers are typically processed in batches of 25 or 50. It is not uncommon to have 10,000 or more circuits fabricated simultaneously. Once complete, the individual circuits on the wafer are cut apart and each one is packaged to form a usable IC. In this section we will describe how the substrates are made.

In the case of silicon, the raw material is an oxide of silicon, called *silica* or *quartz*, depending on the exact structure, in the form of sand. The chemical formula is SiO_2. This is readily available and easy to quarry. The silica sand is *reduced* (the oxygen is removed) in a high-temperature furnace with carbon (coke). The resulting poor-quality silicon is reacted with *hydrogen chloride*, an acid gas, to form compounds that vaporize, such as *trichlorosilane*. This material is easily refined and purified using bulk processes. The removal of unwanted impurities is critical, as they can create imperfections in the final crystal and can alter the electrical characteristics of the semiconductor. The trichlorosilane is then decomposed by heat to form pure silicon *ingots*. The ingots are not single-crystals but are in fact a highly imperfect *polycrystalline* form of silicon. This polycrystalline silicon, although pure, is not the   material for substrates, as it has too many physical imperfections in its structure. Therefore, the polycrystalline ingots are melted and used in the single-crystal formation process.

Compound semiconductor raw materials are more difficult to obtain and are therefore more expensive. For instance, gallium is a by-product of aluminum smelting and arsenic is a by-product of copper smelting.   For compound semiconductors, the material components are brought together in the crystal-forming apparatus after being purified by processes which are somewhat more complex than the refinement processes for silicon.

Crystals of semiconducting material are normally grown in a machine known as a *crystal puller*. Typically, the purified material(s), such as the polycrystalline silicon, is melted by high temperatures in a *crucible*. If the substrates have to be doped, the appropriate type and amount of dopant is added at this stage to the melt. The puller must create a crystal which is as free from lattice defects, imperfections in the crystal structure, as possible. Defects can disrupt the rectifying properties of p-n junctions and hence must be avoided in device-quality material.

Many different crystal growth methods are employed, depending on the type of crystal desired. These are:

Czochralski (CZ). A *seed* crystal, which is a small perfect crystal, is dipped into the melt, slowly rotated, and pulled from the surface. The melted semiconductor crystallizes or "freezes" onto this seed as it rises, to form a larger perfect single crystal extending from the end of the seed to the melt (Fig. 3.7). The growing crystal replicates the crystal *orientation* of the seed. Orientation is an important concept in crystal growth. We may look at a three-dimensional lattice structure from many different angles and will see different atomic arrangements, depending on our view. If we cut a crystal at an angle, we will reveal a particular face or *plane* of the crystal and thereby set the orientation at the surface. Since the growing crystal follows the structure of the seed, the orientation of the seed will determine the atomic arrangement of the surface of the wafer once the ingot is sliced.   Substrate orientation is a factor in MOSFET performance. The act of rotating the seed during pulling results in the formation of a cylindrical ingot or *boule*.

Float Zone (FZ). CZ growth has one drawback: small amounts of oxygen from the air can become incorporated into the melt, creating crystal defects. In the FZ technique, a ready-formed CZ ingot is further refined by locally remelting a zone which travels along the length of the crystal (Fig. 3.8). If we melt part of a crystal in this way, impurities such as oxygen tend to stay in the

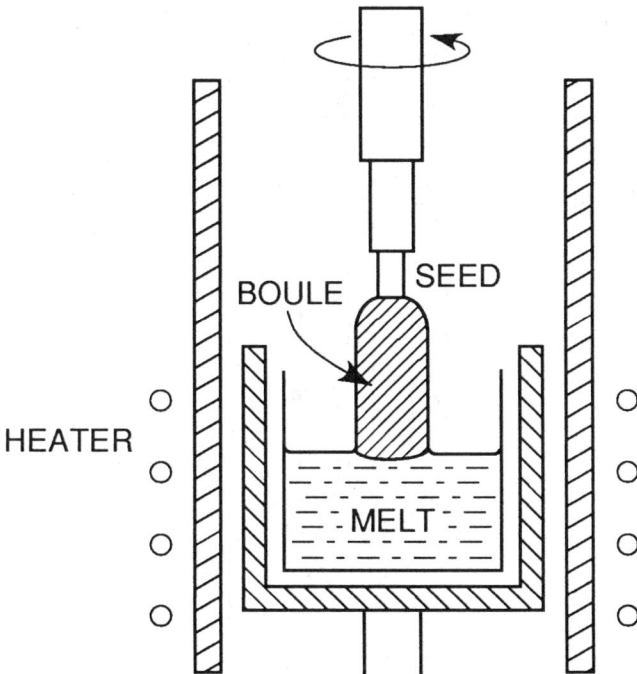

Figure 3.7 CZ crystal puller.

melt region. If the melt region is made to travel down the crystal, the recrystallized part behind the melt is left without impurities.

Liquid Encapsulated CZ (LEC). This is used for compound semiconductor growth in which some of the components are

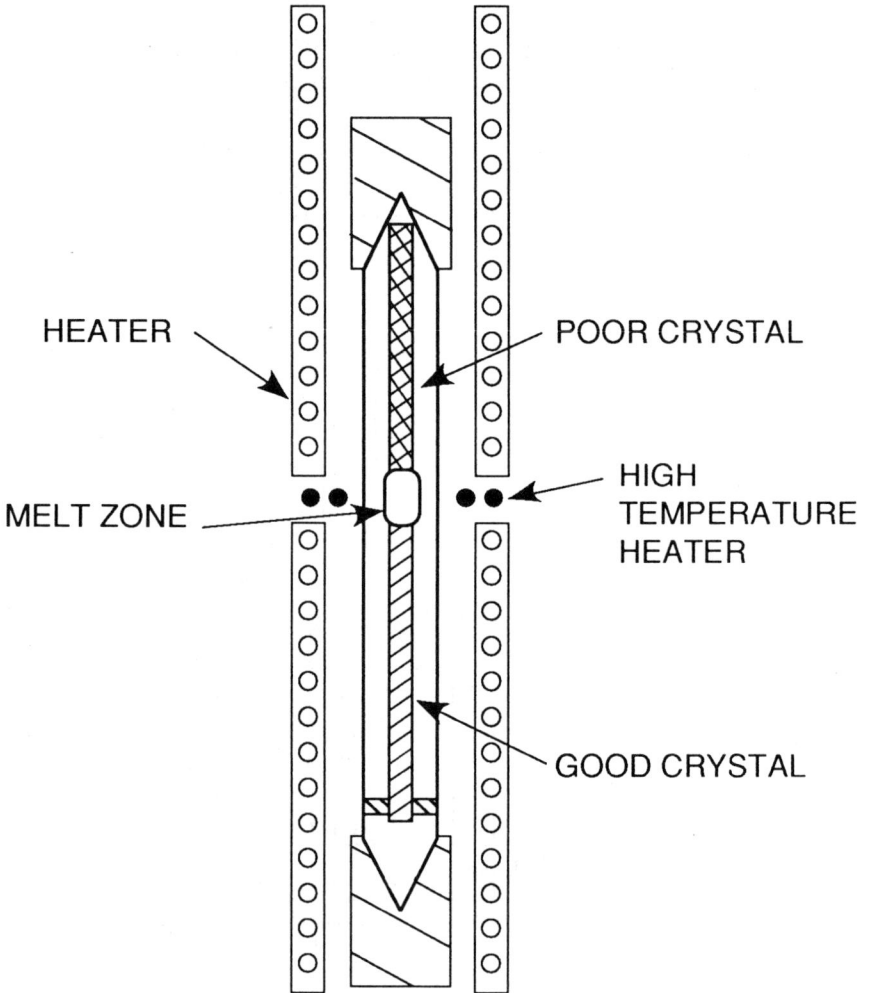

Figure 3.8 FZ refining/pulling system.

volatile; for example, in growing GaAs crystals, we heat arsenic and gallium together, but the arsenic tends to evaporate rapidly when heated. Ordinary CZ would thus result in the loss of component elements through evaporation. A capping layer of molten *boric oxide* is used to prevent evaporation and thereby keep the components of the melt in place (Fig. 3.9).

Bridgman This is a more basic technique for compound semiconductor growth which uses a sealed tube in a moving furnace. The components are placed in a crucible with a seed attached and the furnace is moved past them, melting and recrystallizing them to form a D-shaped ingot (Fig. 3.10). For GaAs growth, additional arsenic is placed in the tube to create an overpressure of arsenic, which reduces evaporation.

Although in most cases the dopant materials such as boron or phosphorus are added when the materials are melted, some silicon ingots are doped by *neutron transmutation doping (NTD)*. This is a very expensive method involving irradiation by neutrons to transform, in a somewhat alchemistic manner, a small number of silicon atoms into phosphorus. The advantage of this technique is that the distribution of the newly created dopant is extremely even throughout the ingot.

The grown cylindrical crystal ingot must now be further processed to be useful in IC manufacture. The ingot is first milled down to the final diameter (e.g., 150 mm), and then *flats* are milled along its length to mark the crystal orientation and the type (p or n). The ingot is then cut up into wafers. This is done using an *annular diamond saw*; this type of saw is mechanically stable and permits the ingot to be cut with great precision. The slice thickness is typically around 0.5 mm for a 100-mm wafer. Since cutting scratches the wafer's surface, the wafers must be *polished*. This process uses a rotating polishing table. Initially, a fine abrasive is used, working down to superfine and then chemical polishing agents. This leaves a surface which is many times better than high-grade optical surfaces with negligible surface roughness. The wafer edge is then rounded to help prevent chipping during handling.

Now that we have our wafers, the next step is to deliberately cause a great deal of damage to the crystal at the back surface of the slice, the surface we will not use for our circuits. The damage is created either mechanically, chemically, or by laser heating. This is

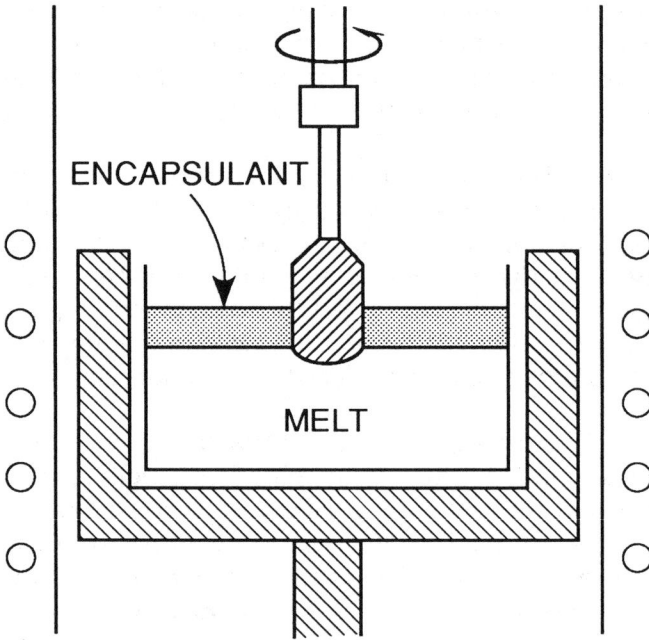

Figure 3.9 LEC crystal puller.

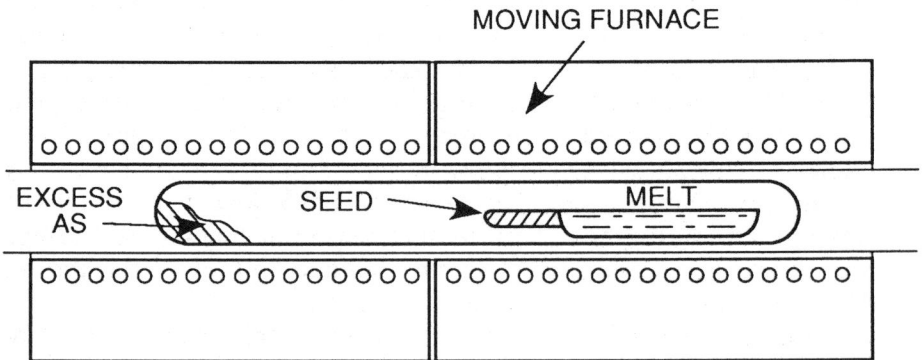

Figure 3.10 Bridgman crystal growth system.

called *backdamage* for obvious reasons, but why it is done is not so obvious. Heavy damage to the crystal at the back of the wafer draws residual defects away from the sensitive front during heating. The defects in the crystal naturally *migrate* to the backdamaged region, as defects like to band together, leaving the surface where the devices will eventually be formed relatively defect free. Another curious thing we should be aware of is that some defects in the wafer *bulk* (away from the surface) are actually beneficial from a mechanical viewpoint. For instance, if we deliberately leave some residual oxygen, a few parts per million, in the bulk, it leads to the formation of defects which help to relieve *stresses* in the wafer during handling and processing. A wafer which has virtually no defects in the bulk is more likely to break when heated quickly or knocked. Therefore wafer manufacture tends to involve a balance of defect/oxygen removal and retention, but in all cases, the device surface should be as free from defects as possible.

So far, we have made a distinction between substrates of silicon and substrates of compound semiconductors such as GaAs. However, both types of substrate fall into the category of *bulk substrates*. A bulk substrate is *homogeneous*, that is, it is uniform throughout. However, there are many other classes of substrates, manufactured for many different technologies, which we may consider. These fall into the following categories:

Epitaxial. An epitaxial substrate is a bulk substrate with a specially grown layer of different doping or even a completely different material put on in such a way that the crystal form of the bulk is continued in the epitaxial layer. In *homoepitaxy* the *epi-layer* is the same material as the bulk, for example, lightly doped silicon on a heavily doped substrate. In *heteroepitaxy* the epi-layer is different, for example, GaAs on silicon. One interesting advantage of epitaxy is that we can put heavily doped *buried layers* in place before we put on the epi-layer. This is useful for some device types.

Insulating. Combining insulating substrates with a semiconductor (or vice versa in some cases) enables electrically isolated islands of semiconducting material to be formed, separated by dielectric. Some circuits require this degree of isolation, particularly if they are to operate at high voltage. A good example of this approach is *silicon on insulator (SOI)* technology, in which the insulating substrate is usually *sapphire,* an insulating oxide of aluminum. Sapphire is particularly good, as nearly single-crystal silicon can be grown on it by heteroepitaxy. *Silicon on sapphire (SOS)* has been produced commercially for some time but is rather expensive and difficult to process (the substrates break very easily due to the stresses created by putting a silicon layer on a sapphire substrate). An alternative to SOS is *dielectric isolation (DI),* in which trenches are etched in a silicon wafer and filled with a suitable dielectric, usually *silicon dioxide,* to provide electrical isolation between areas. In DI, the isolation is usually performed during, rather than prior to, processing. Two other promising technologies are *direct silicon bonding (DSB)* and *implanted oxide.* In DSB, a wafer with a thick silicon dioxide layer on its surface is bonded to another silicon wafer by pressure and heat. The top wafer is then thinned to leave a single-crystal-device quality layer which may be etched away in places to provide isolation. The implanted oxide method creates a *buried oxide* layer underneath a single crystal layer by shooting oxygen into a silicon substrate so that it comes to rest below the surface (we will discuss implantation in Chapter 4).

Summary

We have now seen how discrete devices in their many forms can be connected together to form a circuit using a PWB. However, the more discrete components we attempt to incorporate into a circuit using this method, the less reliable the total circuit or system becomes. The problem may be reduced by integrating most of the components to create circuits which are not only more reliable but also smaller, cheaper, and potentially faster. The semiconductors themselves become the supporting circuit boards in ICs. The materials for these wafers are refined and put into crystal pullers to create single-crystal ingots. These, in turn, are cut and polished to create the wafers. Variations on the substrate theme allow us to create wafers which combine different semiconductors or provide complete isolation for areas within the circuit.

4

The Art of
IC Fabrication

From Designs to Semiconductors

In the last chapter, we saw how we could create the necessary substrates on which to base our devices, circuits, and systems. We will now discuss how we actually process these substrates in order to make ICs. We will assume at this point that the design of the circuit has already been done so that we may move directly to the discussion of the art of semiconductor processing. Design itself will be discussed later in the book. It is acceptable to tackle the subjects in this order, as the *rules* for the designs are based on the capabilities of the fabrication process. We must therefore understand the process and what it can be used to make before we begin to discuss design. In essence, when it actually comes time to make an IC, we first need a working process. We use what we know about this process to create a circuit design, and we then use the design in the process to fabricate our circuit.

Once the circuit has been designed on the design system, a set of *masks* must be manufactured. These masks are the link between the design and the real circuit and will be used in the fabrication process. The process of using masks to transfer designed geometric shapes to the actual materials used to fabricate the IC is called *photolithography*. These shapes represent the physical realization of the circuit, as they will ultimately determine the sizes and positions on the substrate of p-n junctions, dielectric regions, and wiring paths. The process has much in common with the pattern definition technique used for PWB manufacture. How we get from design to masks is a complicated process in IC manufacture, particularly when very small geometries (less than 1 micron) are involved.

The designed circuit is held in the computer as a set of *masking levels*. Each masking level corresponds to a particular layer within the IC. Each of these levels in the computer is used to make a mask which contains all the geometric shapes necessary to form the corresponding level on the finished IC. Each mask is typically used only once in the process. We could have as many as 18 masks in a particularly complex process, although 12 is more common. We will see the full significance of this when we discuss how practical devices are fabricated in the next chapter.

The computer data for each masking level are used to drive a computer-controlled *pattern generator*. This was traditionally an optical machine which could flash out geometric shapes on a photosensitive material called *photoresist*, which covered a *chromium*-coated glass mask. The entire layer could be transferred to the mask using a series of "move-flash" instructions; the mask was moved and exposed by a small rectangle of light, moved again and exposed, etc., in order to trace out the shapes sequentially. Depending on the type of photoresist used (see later), the exposed or unexposed areas became dissolved in an appropriate developer, leaving the areas of chromium, which ultimately had to be removed uncovered. The mask was then dipped in a chromium etch and only the photoresist-covered areas remained, leaving the other areas transparent (uncoated glass).

For small geometry patterns, electron-beam machines have replaced optical systems, as the *resolution* available with *e-beam* is greater. In this case, the computer program is used to deflect the electron beam, steering it to draw shapes on a chromium-coated glass mask which has an electron-sensitive covering on the chromium. When developed, the electron-exposed resist protects the chromium in selected areas during chromium etching and leaves the desired pattern on the mask, as before. The chromium is opaque and therefore will block light well when lithographic transfer is performed.

The pattern generator is generally used to create a *reticle*, which is a large version of the final mask (usually 10x). This reticle may be used in two ways. It may be *photo-reduced* and the small image duplicated to create a *master* mask composed of hundreds of images of the circuit at the final size on a single plate. This process is performed by a *step and repeat system* which projects the image of the layer at its final size on another photoresist/chromium-coated plate. The image is stepped across this plate and is consequently duplicated in a regular array. This is useful, as hundreds of ICs may be fabricated simultaneously using this type of mask. An example of this type of mask is shown in Fig. 4.1. The other way a reticle can be used is directly in a piece of processing equipment called a *direct step-on wafer (DSW)* or *stepper* (see later). In this piece of equipment, the reticle generally contains an image of the layer which is larger than the final image, and the photo-reduction is performed during the transfer process to the wafer.

The chromium or *"chrome"* master mask may be used to mass-produce other masks. These may also be chrome masks, which have the advantage of being hard-wearing and precise, or they may be

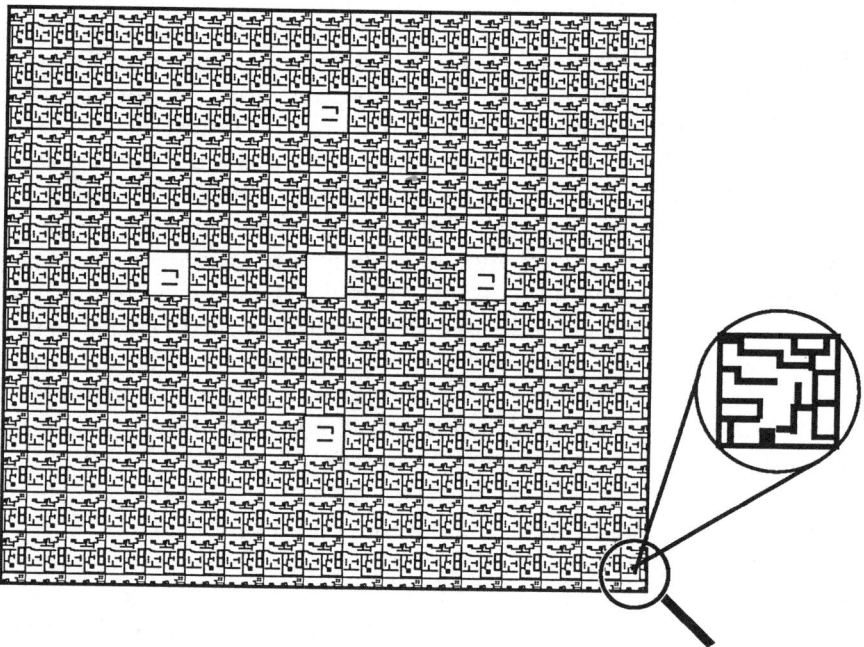

Figure 4.1 Single masking layer.

emulsion masks, which use an emulsion coating instead of chromium and are a low-cost alternative. These are not as hard-wearing or precise as chrome and tend to be used for less demanding applications (very low-volume, low-cost or large-geometry circuits).

Pattern Transfer

Photolithography is a critical process in the fabrication of ICs, as it enables the patterns on the masks to be transferred to the wafers. The production of a typical IC requires multiple photolithographic stages.

   The pattern of the mask is used to pattern the materials on the wafer to form the circuit elements. For example, as we will see later in this chapter, when a metal layer is deposited, it initially covers the entire wafer. Much of the metal must be removed to leave just the interconnecting "wires" which will connect the devices in the circuit together. The pattern or shape of these interconnections is on the appropriate masking layer. The entire photolithography sequence for the example of the metal layer is shown in Fig. 4.2.

   In the photolithography process, a photosensitive material called *photoresist* is deposited on the surface of the wafer. This is performed by a *spin-on* process in which the wafer is spun rapidly and a measured amount of photoresist is dropped onto the surface. Since the photoresist is initially in liquid form (it comes dissolved in a solvent), it is thrown toward the edges and coats the surface evenly in the process. Photoresist is an *organic* (containing *hydrocarbon* compounds) substance which, depending on the chemistry employed, will either be dissolved by an appropriate *developer* after being exposed to light or will be dissolved by the developer if not exposed. The material which is made soluble by light is called *positive resist*, as it forms a positive image of the opaque pattern on the mask in the resist after development. The material which is made insoluble by light is called *negative resist*, as it leaves a negative of the mask pattern after development. After being deposited, the resist is gently heated in a process known as the *soft bake* to drive off the *solvents* which keep the material in liquid form during the spin-on process.

   Optical exposure may be performed on a machine called a *mask aligner* (sometimes also called a *printer*). This places the mask and the photoresist-coated wafer together and shines *ultraviolet light* through the mask. The aligner produces an even illumination over the entire surface of the wafer so that no one part becomes over- or under-

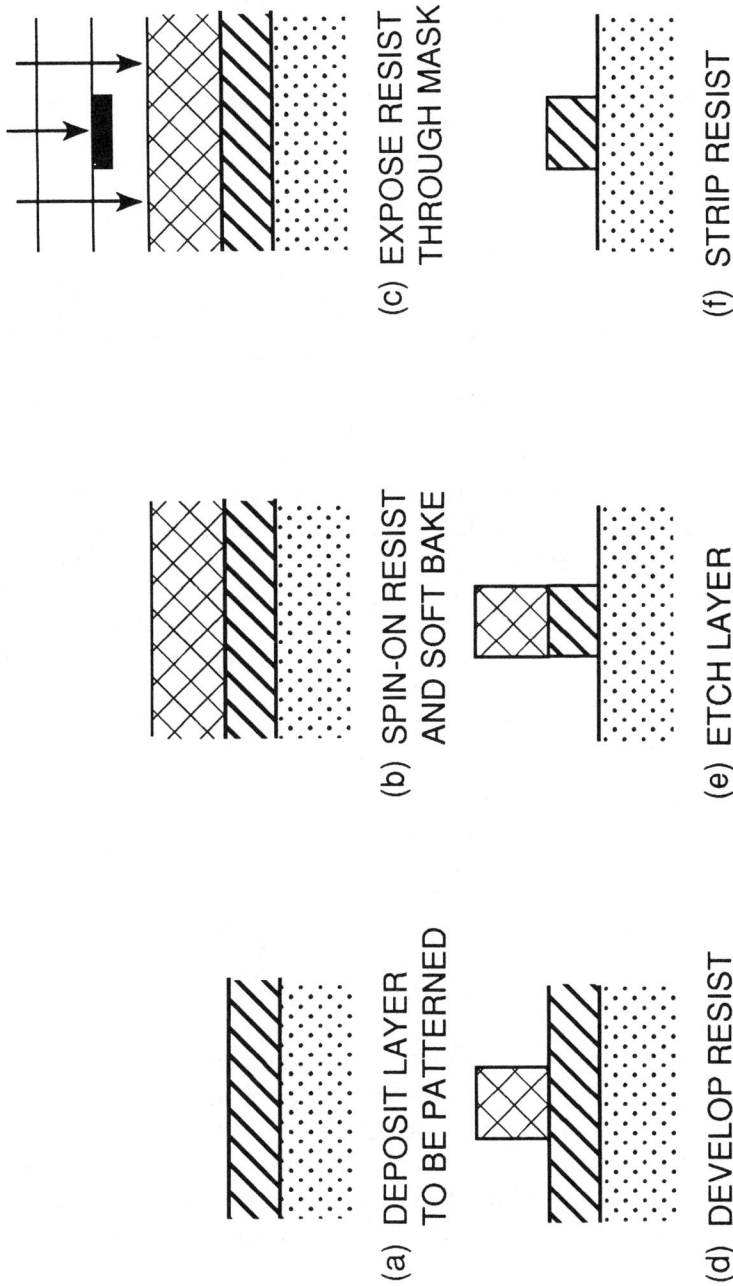

(a) DEPOSIT LAYER TO BE PATTERNED

(b) SPIN-ON RESIST AND SOFT BAKE

(c) EXPOSE RESIST THROUGH MASK

(d) DEVELOP RESIST

(e) ETCH LAYER

(f) STRIP RESIST

Figure 4.2 Patterning a layer by photolithography. (a) Deposit layer to be patterned. (b) Spin-on resist and soft bake. (c) Expost resist through mask. (d) Develop resist. (e) Etch layer. (f) Strip reuse.

exposed. If improper exposure occurs, either the pattern transfer fails altogether or the transferred lines come out with the wrong dimensions (a good analogy is over- or underexposure of a photograph). Photolithography is carried out in a room with yellow lighting, the "yellow room," as the photoresist is least sensitive to these wavelengths.

The exposed wafer is then developed using appropriate chemicals, and a positive or negative version of the pattern on the mask thus appears on the photoresist on the wafer. The resist is then baked to harden it chemically; this step is called the *hard bake*. For the example of the metal interconnection layer, the wafer is placed in an appropriate metal etch to remove the metal not covered by the photoresist. The pattern of the mask is thus transferred to the metal layer, and the wires are formed. Since the photoresist is no longer required, it is now removed or *stripped*. Stripping involves strong acids, solvents, or plasmas (see later) to dissolve the photoresist layer. This entire process is repeated many times during device manufacture.

A single mask set, the set of masks of all layers required to make up a particular circuit, may be used to manufacture hundreds of thousands of devices. If a *contact* process is used in which mask and wafer are in hard contact during exposure, then the masks will eventually become scratched and worn down. The alternative is a *proximity* process in which a very small gap is maintained between the mask and wafer. The gap cannot be too large; otherwise, the image which reaches the wafer will become blurred by *diffraction* of the light. This cuts down on mask wear and contamination from the photoresist, but the mask and wafer are still close enough so that a dust particle can damage both mask and wafer. A better technique in this respect is *projection* lithography, whereby the mask and wafer may be a few feet apart. Complex and expensive optics ensure that the projected image reaching the wafer is not distorted in any way.

The term *mask aligner* (or *contact aligner*, *proximity aligner*, or *projection aligner*) is used, because the layers of the devices and circuit must be accurately matched up or *aligned* with one another for the circuit to work properly. For instance, if a metal (wiring) layer is badly misaligned, the ends of the connections may not meet the devices and the circuit could not operate. This is an extremely accurate process, as the layers often have to be matched to an *overlay* or *lateral tolerance* of less than a few tenths of a micron for optimal circuit performance. Special marks called *alignment marks* are placed on every masking level to help the alignment process. The alignment mark on the mask and the alignment mark from the previous layer, now defined on the wafer by the previous process steps, are aligned either by an operator using a microscope on the aligner or by automatic means.

For very small geometry work and large wafer diameters (150 mm or greater), a large mask containing hundreds of identical circuit patterns will cause problems, as the optics and the mechanical systems of the aligner cannot be made well enough to prevent uneven exposure or image distortion for small features over such a large area. A DSW or stepper uses a reticle at 10x, 5x or 1x of the final circuit size and steps the image at final size directly on the wafer. Since only one circuit or a small number of circuits (two to eight, depending on size) are exposed at a time, the final image quality may be considerably better. It is much easier to align the circuits one by one than to try to align several hundred simultaneously over a large area to a very small overlay tolerance. The alignment is performed automatically on these systems. Also, since the reticle can be 10x or 5x the final size, a 1 micron line (final size) can be as much as 10 microns on the reticle; hence the reticles are easier to make.

One of the most significant factors limiting the resolution of lithography systems is connected with the source of energy used to expose the resist. We know from our earlier discussion that ultraviolet light is used in optical systems, but the process is not quite so simple. The source of the light is generally a *mercury vapor* lamp which produces not one but several *wavelengths* of ultraviolet light, which emerge as strong peaks in the intensity from the lamp. We label these wavelengths with letters; for instance, the 0.436 micron peak or *line*, as it is called, is labeled the *G-line*, whereas the shorter wavelength 0.365 micron peak is labeled the *I-line*. If this is beginning to sound like a physics lesson, do not fear because one very simple fact emerges: shorter wavelength light can resolve smaller features. Therefore, an I-line stepper has better resolution capability than a G-line system. Of course, the photoresists, the aligner optics (lenses, mirrors, etc.), and, to some extent, the mechanical systems have to be different in both cases.

Unfortunately, even the I-line becomes unsuitable for extremely small features, those less than 0.5 micron in size. Therefore, the trend in advanced lithography is to move away from mercury lamps to other shorter wavelength sources, particularly in the fabrication of such high-density products as *dynamic random access memories*

(DRAMs).　One such source is the *excimer laser*, which is able to produce very short wavelength ultraviolet light.　Better still are *x-ray* sources, which have effective wavelengths of a few thousandths of a micron.　These x-ray systems would be the ultimate lithography tools if it were not for the problems of using x-rays.　For instance, x-ray masks are difficult to fabricate and x-ray "optics" basically do not exist.　The most advanced x-ray systems in use today get the necessary intensity, uniformity, and area of exposure by the use of a *synchrotron* x-ray source.　This is an extremely expensive method as a synchrotron is effectively a large particle accelerator costing tens of millions of dollars.　Therefore, one synchrotron typically has a number of exposure units attached to it.　Another alternative to optical exposure is *direct write electron-beam* exposure.　No masks or reticles are used, as the pattern is written directly by the beam on electron-sensitive resist on the wafer.　The advantage of direct write techniques is that extremely small linewidth circuits may be fabricated due to the small effective wavelengths involved.　It is also an ideal technique for prototype and research work, as the design program is used directly to steer the beam; therefore, delays in producing masks are eliminated.　The alternative direct write technique uses a *focused ion beam (FIB)* rather than an e-beam.　This has some interesting advantages, including the ability to use dopant ions in the beam and make small geometry p-n junctions just by moving the beam to wherever the junctions have to be formed; no resist is necessary.　Unfortunately, direct write techniques have a severe drawback.　Since every line on every circuit has to be drawn one at a time, the technique is very slow compared to lithography systems, which expose the entire circuit layer with one flash.

Depositing Layers

The fabrication of an IC requires many processes involving the *deposition* of various materials.　*Thin film* deposition involves the transfer of a material from a deposition *source* to the surface of the wafer.　It differs from the *growth* of a material (see later) in that there is essentially little reaction between the material and the substrate; we put the thin film on the surface rather than making a layer by reacting something with the surface.　In this section, we will discuss deposition and introduce some of the common materials used in IC manufacture.

Deposited layers can be insulators/dielectrics, conductors, or semiconductors. The deposited materials may ultimately be patterned by photolithography and an appropriate etching step using an etchant which will remove the material but leave the photoresist intact. There are two main categories into which we can group deposition processes: *physical vapor deposition (PVD)* and *chemical vapor deposition (CVD)*.

One form of PVD is called *evaporation*, in which the source of the material to be deposited is heated so that it evaporates and subsequently recondenses on the wafers. An alternative form of PVD, called *sputtering*, uses a stream of *ions* to bombard a *target* made of the material to be deposited. Both of these techniques are shown schematically in Figs. 4.3 and 4.4. The bombardment dislodges material from the surface of the target. This material lands on the wafers to form a thin film.

In CVD, the source of the material to be deposited is already in gas or vapor form. It is *decomposed* by heat or another source of energy (radio frequency power, light, etc.) after it lands on the surface

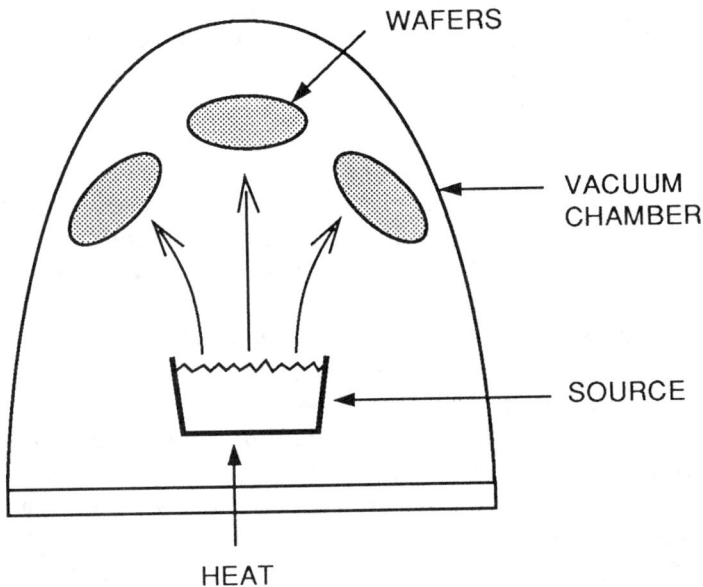

Figure 4.3 Evaporation system.

of the wafer, leaving behind a thin film of the desired material. Multiple gases may be mixed in the process chamber and simultaneously decomposed to form films with complex compositions. A typical CVD system is shown in Fig. 4.5. Both types of deposition are performed after the air has been flushed from the process chamber so that the component gases of the air (mainly nitrogen and oxygen) do not interfere with the process. We will now examine the most common materials deposited by PVD and CVD and elaborate on the processes for specific materials.

Semiconductors

Silicon. When we discussed the different types of substrates in Chapter 3, we mentioned *epitaxial layers* in which a thin layer of semiconductor is deposited on a substrate of silicon, sapphire, etc. The process of silicon epitaxy is actually a CVD process in which a silicon-containing gas such as *trichlorosilane* ($SiHCl_3$) is decomposed at high temperature (about 1100° C) to form an epitaxial layer on the surface of a compatible substrate. The epi-layer may be doped during deposition by adding a dopant-containing gas such as *diborane* (B_2H_6) for *boron* doping, *phosphine* (PH_3) for *phosphorus* doping, *arsine* (AsH_3) for *arsenic* doping and *stibene* (SbH_3) for *antimony* doping. The process is generally carried out at *atmospheric pressure* (the same pressure as the surrounding air); such processes fall into the category of atmospheric pressure CVD or APCVD.

Gallium arsenide. Epitaxy for GaAs and other compound semiconductors is currently performed using highly specialized forms of PVD and CVD. The difficulty with compound semiconductors is that not only do parameters, such as thickness, have to be controlled during deposition, but strict attention must be paid to the precise composition and crystalline form of the layers. *Molecular beam epitaxy (MBE)* uses an ultra-high vacuum system and a series of sources which evaporate the material to be deposited under highly controlled conditions. It is possible to deposit gallium and arsenic together to form extremely thin GaAs layers or to include a third material, for example a dopant, in these layers. Layers a few atoms thick have been produced using this technique. The CVD alternative is a technique called *metal organic CVD (MOCVD)*, alternatively called *organometallic CVD (OMCVD)*. As the name suggests, it is a type of CVD which creates thin layers of GaAs or other compounds (although

POWER
SUPPLY

VACUUM
CHAMBER

TARGET

ION
MOTION

WAFERS

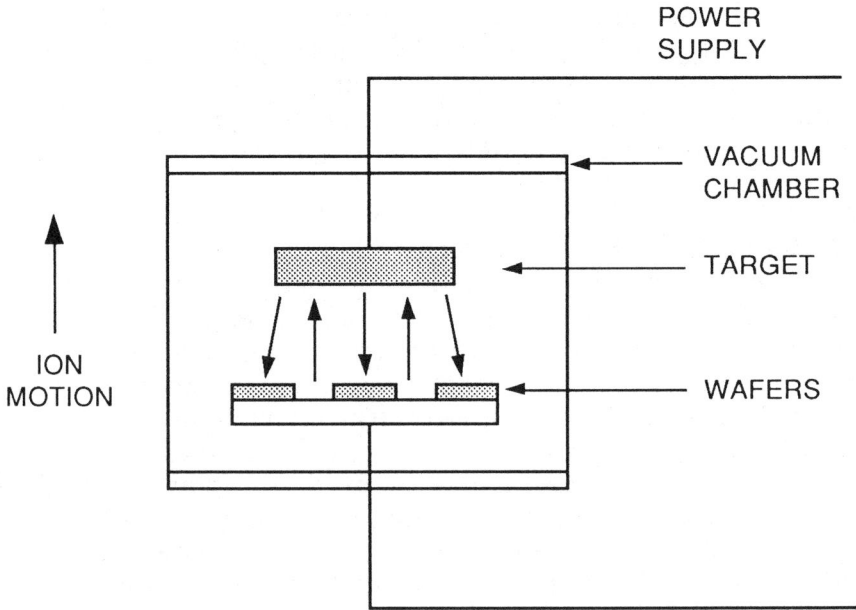

Figure 4.4 Sputtering system.

not as thin as MBE) through the thermal decomposition of organometallic vapors (such as *dimethyl zinc* and *trimethyl gallium*). This technique is not as highly controlled as MBE but is currently better suited for industrial mass production. A relatively new technique has emerged which combines organometallic sources with MBE; it is called MOMBE (for obvious reasons).

Conductors

Polycrystalline silicon. Although it may seem that this should be in the semiconductor category, the polycrystalline form of silicon is not a particularly good semiconductor for device applications, as it contains too many irregularities and *faults*. Heavily doped *poly-Si*, however, is frequently used as a *gate electrode* in MOS devices for reasons we will discuss in the next chapter. The heavy doping makes it a reasonable conductor, so it may be used in MOS gate electrodes and in interconnect applications. Poly-Si is deposited at *low pressure* at about 600° C by the thermal decomposition of *silane* (SiH_2). [The schematic diagram of the representative CVD system in Fig. 4.5 is actually a typical low-pressure CVD (LPCVD) system.]   At this temperature the silane breaks up, leaving silicon in polycrystalline form on the surface of the wafers. The poly-Si may be doped *in situ* (during deposition), by adding appropriate gases, or afterward by *diffusion* or *ion implantation*.

Silicides. The particular *silicides* used in ICs are usually compounds of silicon and *transition metals*. The two most common are *titanium silicide* ($TiSi_2$) and *tungsten silicide* (WSi_2), although others such as *molybdenum silicide* ($MoSi_2$), *tantalum silicide* ($TaSi_2$) and *platinum silicide* (PtSi) may also be used. These materials are good conductors (much better than doped poly-Si but worse than aluminum) and are able to withstand high-temperature processing (which aluminum can not). They are typically used, along with doped poly-Si or doped source-drain regions in MOS devices, in applications such as high-speed ICs which demand low interconnect resistivities at the gate/device level. They may be deposited by a number of methods. The metal component may be evaporated by means of a hot filament onto the surface of the silicon or poly-Si and then heat treated or *sintered* to form the silicide or *polycide* in the case of poly-Si. A polycide is merely poly-Si with a capping layer of silicide to lower the resistance of the structure. Alternatively, the metal component or even the complete silicide may be sputtered onto the wafer. Sputtering involves bombarding a target of the material to be deposited with argon ions and is more controllable than evaporation, particularly when more than one element is in the target. A further method of silicide deposition is by LPCVD using reactants such as silane and *tungsten hexafluoride* (WF_6).

Aluminum and its alloys. *Aluminum* is a widely used conductor in ICs and may be deposited readily by evaporation or sputtering. It is used as a gate material in the older-style *metal gate* MOS devices and as a final level of interconnect in modern circuits. It has an extremely low resistance but cannot withstand high-temperature processing. Therefore, it can generally be employed only later on in the IC fabrication process when all the high-temperature steps have been completed. Aluminum is frequently alloyed with small amounts of other materials to improve its properties. For instance, 1 percent of

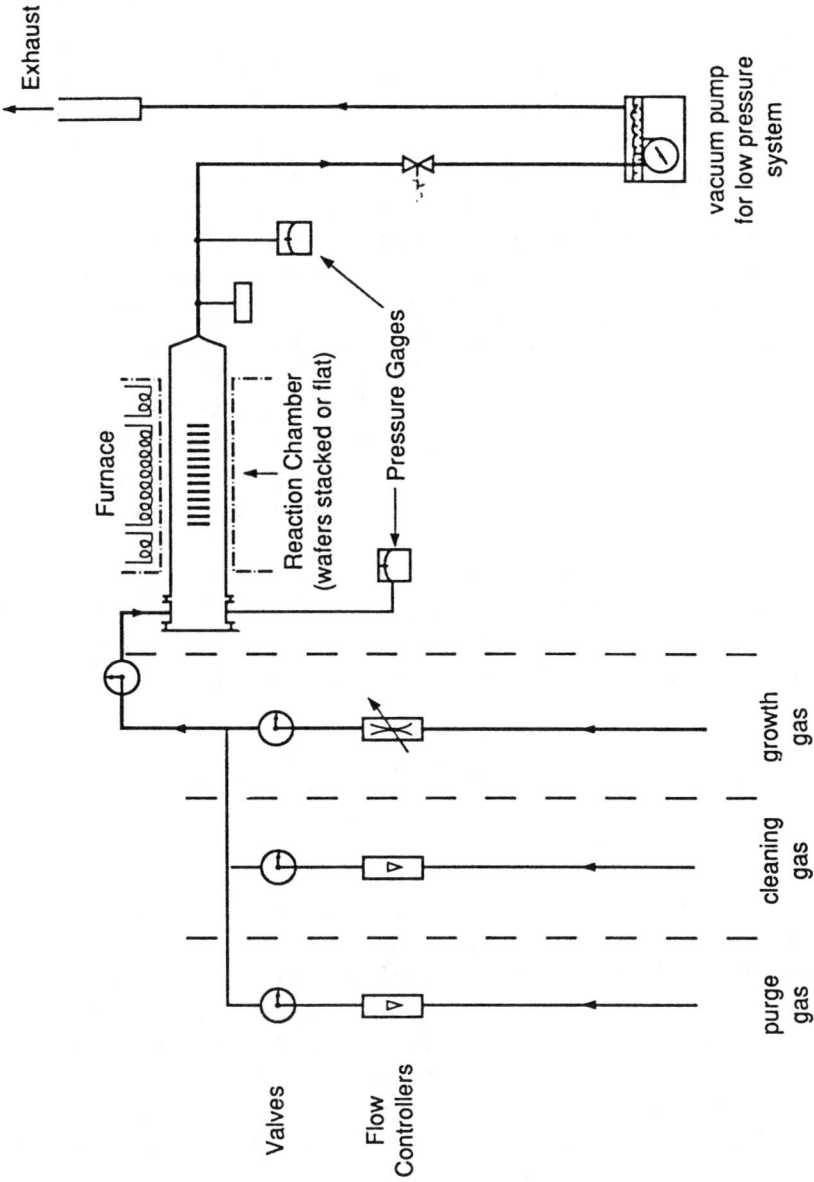

Figure 4.5 Typical CVD system.

silicon may be added to prevent the aluminum from dissolving the silicon substrate (silicon is soluble in aluminum).   Also, a small percentage of copper may be added to harden the metal and make it less susceptible to breakage under high electrical stress conditions.

Refractory metals.   These metals, which include *tungsten,* can be sputtered but are more usually deposited by LPCVD, using *tungsten hexafluoride* for tungsten deposition.   *Refractory* refers to their ability to withstand high-temperature processing.   Tungsten, in particular, has replaced aluminum in many interconnection applications because of this property.   Tungsten is also a rather curious material in that it may be *selectively deposited.*   Under certain conditions, it likes to deposit on silicon but is reluctant to deposit on silicon dioxide, as the oxide prevents the decomposition of the reactants used to form the tungsten layer.   This is important, as we can make the metal fill up deep *contact holes* (connections between the silicon substrate or poly-Si and the overlying metal layer) or *vias* (connections between two layers of metal) in *silicon dioxide* dielectric.   This improves our ability to make high-reliability, low-resistance connections between layers.

Barrier materials.   In many cases, particularly when many different materials are used in interconnection systems within an IC, we do not want dissimilar materials to come in contact because of the chance of a *metallurgical reaction* which could eventually make the circuit fail. We therefore use *barrier materials,* which are conductors that do not allow the intermixing of the materials they separate.   Examples of these materials are *titanium nitride* (TiN) and *titanium-tungsten* mixture.

Dielectrics

Silicon dioxide.   When used as the insulator between the gate electrode and the semiconductor in MOS devices, *silicon dioxide* is normally grown rather than deposited, as grown oxide has better dielectric properties than deposited SiO_2. However, it is much quicker to form thick layers of silicon dioxide for less critical applications by deposition, and the temperatures used are generally much lower than growth temperatures.   One such application is as the insulator between devices in the semiconductor substrate and the first level of aluminum interconnect. SiO_2 is frequently deposited by CVD at about 400° C and at atmospheric pressure using *silane* and *oxygen. Low-temperature oxidation (LTO),* a process similar to LPCVD, and *plasma-*

enhanced CVD (PECVD) are often used to deposit layers of SiO_2 to insulate levels of wiring from each other or to *passivate* (protect) the circuit from the external environment. PECVD uses a high-frequency (radio frequency or rf) electric field to break up the molecules of reactant vapors and gases so that the oxide may be deposited at around 200° C, which is a very low-temperature for deposition. Low temperature deposition (less than 400° C) is important when aluminum is being covered, as the metal cannot withstand higher temperatures.

Silicon dioxide films with a significant content of phosphorus (*phosphosilicate glass* or PSG), boron (*borosilicate glass* or BSG), or both (*borophosphosilicate glass* or BPSG) will protect the circuit better than undoped material against unwanted impurities from the environment. These materials will also *reflow* or smooth out at a lower temperature than undoped glass. This makes it easier to deposit further conducting layers on top, as sharp edges in the oxide can create breaks in an overlying metal film. Another way of producing a smoother film is to sputter the SiO_2 from an oxide target while directing some of the sputtering ions onto the depositing layer. This *bias-sputtered quartz (BSQ)* approach has the effect of smoothing out any peaks in the layer caused by features in the underlying substrate. This creation of a smooth layer which has few vertical steps is called *planarization*. Finally, one method of producing planarized oxide layers is to abandon PVD and CVD and to use *spin-on glass*. In this case, the SiO_2 is contained within a compound which may be dissolved in an appropriate solvent. In the liquid form, the SOG can be deposited onto the surface by spinning the wafer and dropping on a measured amount of material which spreads out to cover the entire surface (much like the photoresist deposition process). Once *cured* by heat, only silicon dioxide is left in the layer.

Silicon nitride. Another commonly used dielectric is *silicon nitride* (Si_3N_4). It can be used as part of a gate dielectric structure particularly in special memory devices, as a passivation layer, or as an antioxidation layer (see later). Since it cannot be grown easily, it is deposited by chemical means.

Polyimide. All the materials we have discussed so far (with the exception of photoresist) are *inorganic*, that is, not consisting of *hydrocarbon-type* compounds. *Polyimide* is an *organic* dielectric which cannot be deposited by PVD or CVD, as it would not survive these processes. It is actually deposited by the same spin-on techniques used to deposit photoresist. It is used as an insulator between metal layers and as a passivator. It does not require reflow, as the spin-on process produces a naturally smooth covering, and it is deposited at room temperature, so sensitive metals do not suffer.

In general, CVD processes produce films which are more *conformal* than PVD. This means that the deposited layers cover the features on the wafer more evenly. This even coverage becomes particularly important when we have steps or holes in a previously formed layer on the substrate. For instance, if we have a steep step and use evaporation to deposit the metal over it, there is a risk that the metal will be thin at the bottom of the step due to the physical processes occurring. This could lead to circuit failure at some point in its operating life. The *step coverage* for CVD is much better, but the situation can be improved for PVD by reflow or planarization of the underlying layer, or by using substrate heating while depositing a metal layer to help smooth it out as it goes down.

Growing Layers

Oxidation is the *growth* of an *oxide* on the surface of a material. Growth differs from deposition in that there is a chemical reaction between the oxidizing ambient (*hot oxygen gas—*O_2, or *steam—*H_2O) and the substrate. Some substrate material is consumed in the process of forming the oxide; hence the oxide layer will typically extend above and below the original substrate surface level. In the case of silicon, the oxide is, of course, silicon dioxide. The silicon from the silicon substrate combines with the oxygen in the oxidizing ambient to form the SiO_2. This requires temperatures in excess of 800° C for the reaction to take place at a reasonable rate. *Thermal oxidation* is unsuitable for many compound semiconductors, as the substrates decompose at high temperatures in oxidizing ambients. Thermally grown SiO_2 is an excellent gate insulator in MOS technology. It is better than deposited oxide, as it contains fewer *incomplete bonds*. Such breaks in the atomic structure will lead to *trapped charge* in the oxide which is undesirable for high-performance MOSFETs.

Oxidation is performed in an *oxidation furnace*. In semiconductor processing, a furnace consists of a *quartz* (a misnomer, as the material is actually fused *silica*) tube within a multiple-zone, electrically heated unit, as shown in Fig. 4.6. Quartz is used for the tube and all

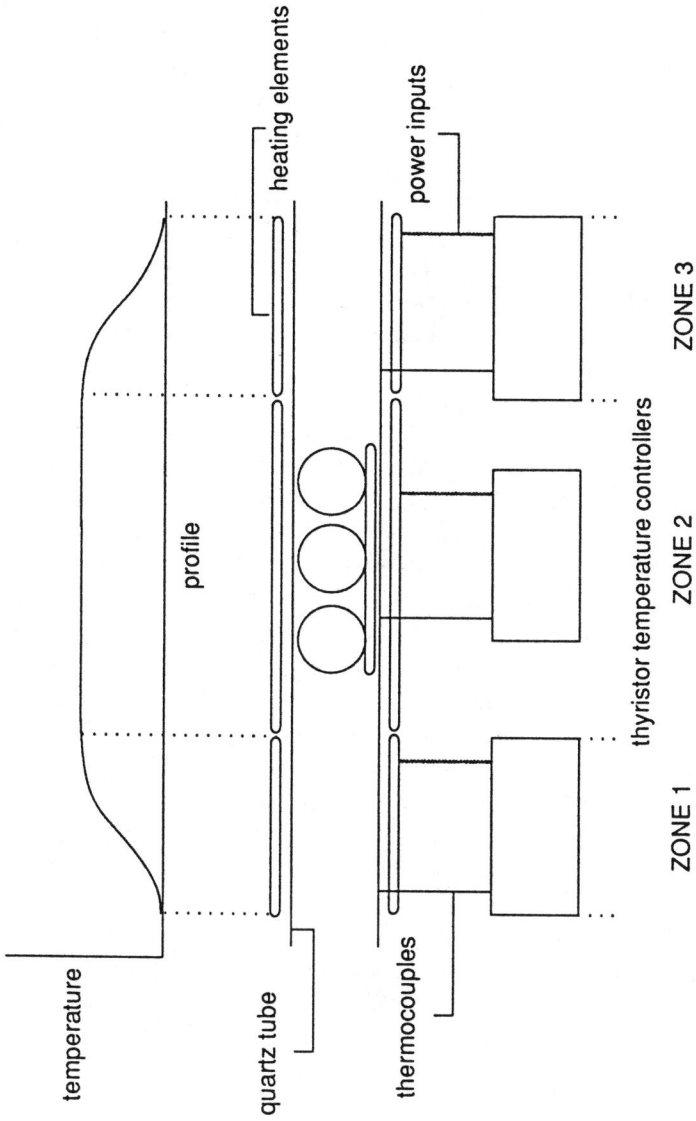

Figure 4.6 Schematic of a diffusion/oxidation furnace

"furnaceware" (end cap, substrate loader, substrate carriers or *boats*) due to its high purity and high thermal stability. Another material which is frequently used for the tube and for components such as the paddle which holds the boats full of substrates is *silicon carbide* (SiC). This material is also very stable at high temperatures. The furnace temperature may be held to better than 0.1 percent by the control system. The oxidation gases enter at one end and are vented at the other. This furnace arrangement is similar to that used in many other thermal processes.

For high-quality oxides which are typically grown fairly slowly, pure oxygen may be used. For fast-grown oxides, steam is a more rapid oxidizer. The steam is formed by bubbling a carrier gas (usually nitrogen) through ultra-pure water or by "burning" hydrogen and oxygen in the tube to create pure H_2O. Oxide growth rate for both cases increases with temperature and time.

Besides being used as *gate dielectrics* in MOS devices, thermally grown oxides may be used as *field dielectrics*. These field oxides support the interconnecting wires in areas not populated by devices and are generally much thicker than gate oxides to reduce the possibility of *parasitic device* formation (the thicker the oxide, the less chance there is of accidentally forming an MOS transistor between the real devices). Very thin oxides are also used in nonvolatile memories to pass charge into the isolated storage areas. We will return to memory devices in Chapter 6. Another important use of silicon dioxide is as a *diffusion mask*. Although dopant materials can *diffuse* into semiconductors, they are less able to pass through SiO_2. Hence the material provides an excellent means of preventing doping in areas where we do not want it. For instance, if we want to form a small p-n junction in a substrate, we can grow an oxide on the entire surface, use photolithography to open up a small window in a photoresist mask, and then use an appropriate etchant to remove the area of oxide not covered by the resist. The photoresist is then removed, and the substrate is doped using diffusion techniques. We will discuss etching in the next section and diffusion in the following one.

Patterning Layers by Etching

Etching involves the removal of a material by *chemical* or *physical* processes. The choice of *etchant* is critical, as it must be as *selective* as possible; that is, it must etch the material to be removed much more rapidly than it attacks the photoresist mask on the surface. The etching process should also effectively stop when the layer to be removed has been etched through. Therefore, the etchant should be selective in etching the layer to be etched while not attacking the substrate or underlying layers.

Wet chemical etching typically uses *acids* and acid mixtures to etch the materials. In the case of silicon dioxide, the etchant is *hydrofluoric acid* (HF), usually *buffered* with *ammonium fluoride* (NH_4F), which has the effect of preventing damage to the photoresist. HF will etch the oxide without appreciable etching of the silicon substrate. Various other materials can be wet etched using a variety of mixtures. For example, one wet etch used for aluminum is a mixture of *phosphoric acid* and *acetic acid*. *Nitric acid* can be used to *strip* (totally remove) photoresist without attacking the underlying silicon or oxide. One drawback of wet etching is that once a chemical has been used, it then contains matter from the etched substrates, and there is a risk of contamination of subsequent wafers unless a spray system is used rather than an etching tank. Also, wet etching is not highly controllable, as it depends on many factors, such as temperature, time, and strength of the mixture. Surface tension of the liquid also prevents small features (holes and trenches), from being etched. Therefore, wet processing is unsuitable for very small geometries.

The alternative to wet etching is *plasma etching*. This technique uses no wet chemicals but a "dry" *gas plasma* instead. (Actually, the term *plasma* is something of a misnomer, as what we really create is a *glow discharge*). The plasma is formed by exciting a gas at low pressure with a high-power radio frequency electric field (Fig. 4.7). This pulls the gas molecules apart and the resulting fractions, called *radicals*, are extremely reactive, more so than strong acids in many cases. There are essentially two types of plasma etcher: a *barrel* etcher, which is used for noncritical work and can take large loads of stacked wafers, and a *parallel plate* etcher, which is used for precision work but generally has a smaller wafer load. The gases used may be fairly inert in gas form. For instance, *carbon tetrafluoride* (CF_4), also called *Freon 14* (a Dupont trade name), is used to etch polysilicon but is sufficiently reactive only in plasma form. Other etch gases can be hazardous; *chlorine* (Cl_2) and chlorine-containing compounds are used to etch oxides and metals. Since plasma etching is a dry technique, there are fewer contamination problems than with wet etching, and smaller geometries may be etched with greater precision, as the process is more controllable (i.e., the etching stops when the power is switched off). It is also possible to have *endpoint detection*, an optical or analytical means of determining when the layer is etched through. An oxygen plasma may be used to strip resist in a barrel etcher in a process known as *ashing*.

One major problem with wet or plasma etching is that the etching process tends to be *isotropic*. This means that the etch process not only proceeds down into the layer to be etched but also moves laterally under the resist (Fig. 4.8). This is called *lateral etching* or *undercutting* and can severely reduce the ability of these techniques to etch small features, especially if steep *sidewalls* are required (which is

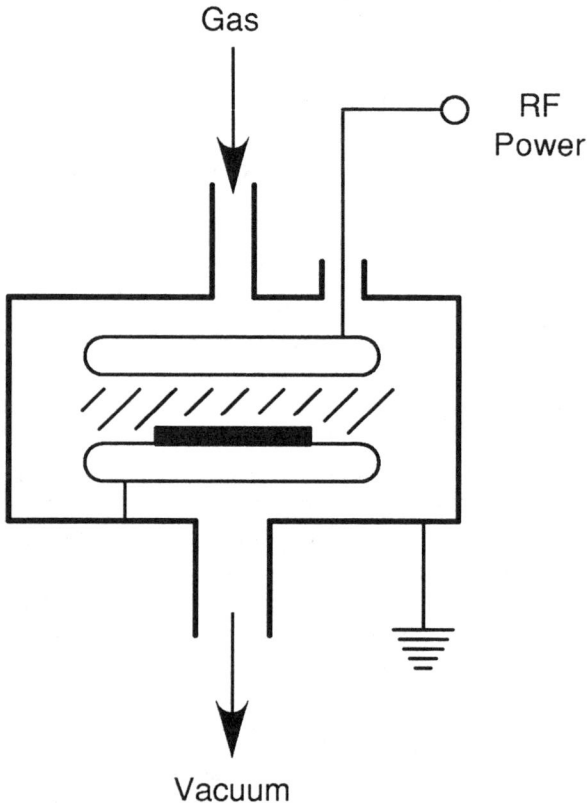

Figure 4.7 Plasma etching system.

often the case in circuits which require small linewidths to attain high *packing densities*).

For cases where *anisotropic* etching is required, in which the etching proceeds more rapidly in the vertical direction than horizontally, a technique called *reactive ion etching* (RIE) may be used. RIE employs a constant dc voltage in the etching chamber as well as the radio frequency field. The radio frequency still creates a plasma which chemically etches the material, but the dc voltage adds a *physical* component to the etch.   This means that the charged components of the plasma, the ions, are accelerated toward the substrate in the vertical direction, which helps to create the anisotropic effect.   RIE uses much lower gas pressures than conventional plasma etching so that the ions can travel toward the substrate more freely.

A fully physical etching method is ion milling.   This uses an accelerated broad beam of argon ions to physically bombard the surface of the wafer. The surface layers of atoms are knocked off, and hence etching occurs.   This method is extremely anisotropic, but photoresist alone cannot withstand the etch.   Therefore, a metal or sputtered oxide surface mask is usually necessary.   This is put on using *trilevel resist* technology, which uses a first layer of resist, a metal or oxide coating on the resist, and then a top layer of resist. The pattern is transferred to the top resist. This is used to pattern the metal or oxide. The pattern on the metal is used as a mask to pattern the underlying resist.   The top resist is removed and the etch performed. The metal may then be easily removed by dissolving the underlying resist.

Creating p-n Junctions by Diffusion

Diffusion is one way dopants may be introduced into a semiconductor in order to form p-n junctions for devices or to change the concentration of dopant in the substrate.   At high temperatures— greater than 400° C for compound semiconductors or greater than 800° C for silicon—a dopant element will *diffuse* or move through the solid semiconductor from a region of high dopant concentration to a region of low concentration. Therefore, to *dope* a semiconductor, the wafers are placed in a *diffusion furnace* at high temperatures in an ambient containing the desired dopant.   Diffusion furnaces are very similar to oxidation furnaces.   The dopant goes from the gas ambient

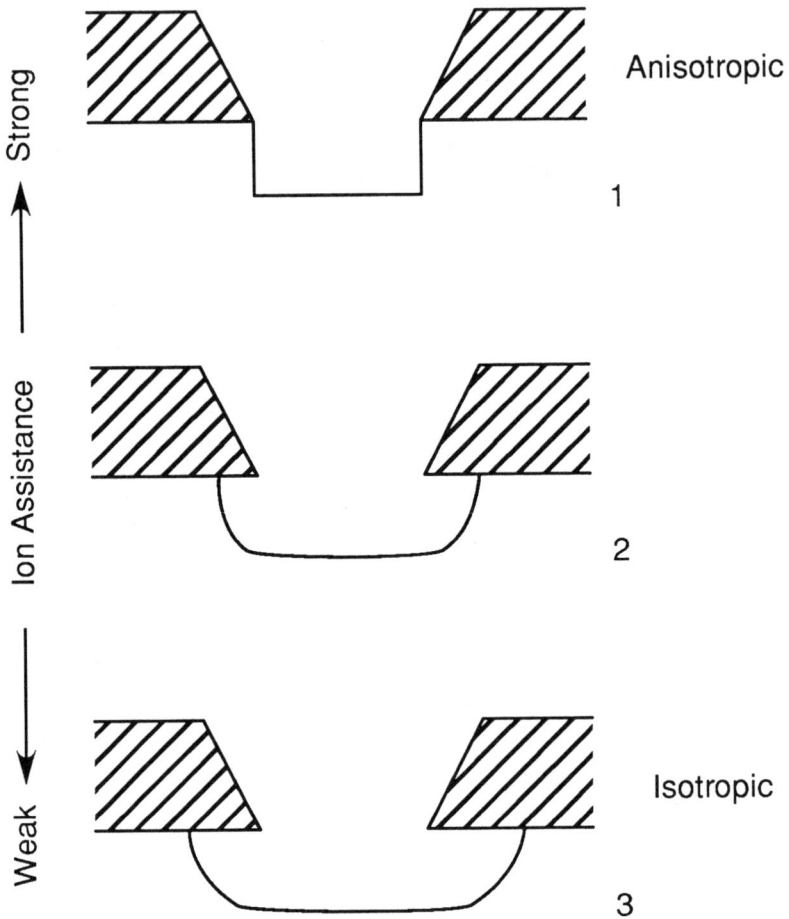

Figure 4.8 Anisotropic and isotropic etching.

surrounding the wafers into an intermediate solid form, typically a doped oxide on the surface of the wafers, and subsequently diffuses into the semiconductor. Dopant sources may be solids placed in close proximity to the wafers, liquids through which a carrier gas is bubbled to collect vapor, or gases which are fed directly into the furnace tube. Examples of solid and liquid diffusion source systems are shown in Figs. 4.9 and 4.10, respectively. Modern *solid sources* for silicon diffusion are discs containing the dopant materials which are loaded into the furnace along with the wafers. These discs contain *phosphorus pentoxide* (P_2O_5) for phosphorus doping and *boric oxide* (B_2O_3) for boron doping. These dopant oxides evaporate and recondense on the wafers to act as solid surface dopant sources. For compound semiconductors, the diffusion tubes are generally sealed off and evacuated to prevent accidental oxidation and decomposition of the substrates, and the solid dopant source is usually contained in a trough at one end of the sealed system. Liquid sources include *phosphorus oxychloride* ($POCl_3$) and *boron tribromide* (BBr_3). Gas sources

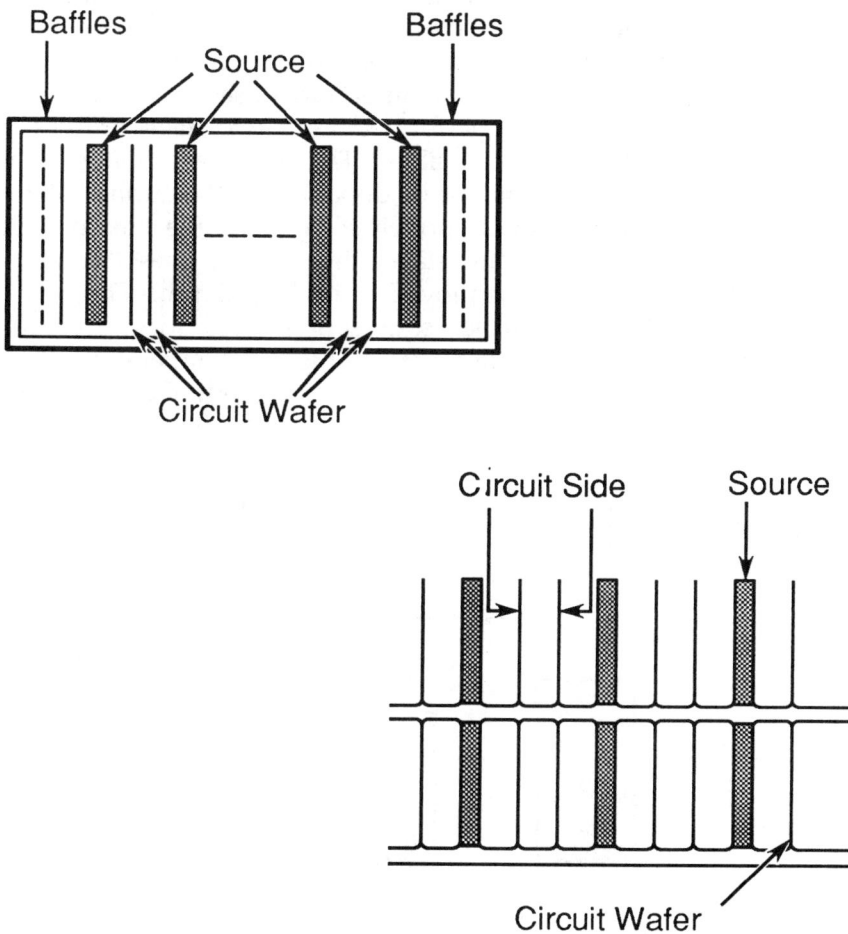

Figure 4.9 Solid source diffusion method.

are generally extremely hazardous metal hydrides such as *diborane* (B_2H_6) and *phosphine* (PH_3). Oxygen gas is added to the hydride gas flow, so that a doped oxide forms on the surface of the wafer and acts as the surface dopant source. Note that this oxide has to be removed after the diffusion process is complete by dipping in HF. The process is called *deglaze*.

Diffusion is used to form the source-drain areas in MOS devices and in the emitter, collector, and base regions in bipolar devices. To prevent doping in areas which must not be doped, an oxide is patterned by the method described in the photolithography section using hydrofluoric acid as an etchant to remove the oxide where not protected by the developed photoresist pattern. It is important to note that photoresist alone cannot be used as a diffusion mask, as it would be destroyed by the high temperatures used in the diffusion process. Diffusion is often performed as a two-stage process to achieve as much control as possible. The first stage involves the *predeposition* of a known amount of dopant into the surface of the wafers, using the doped oxide principle to create the intermediate surface source. In the second stage, this dopant goes through a *drive-in* process in which it is moved to its final depth. Drive-in is performed without any further source of dopant (deglazing has already been performed) and at a higher temperature than predeposition. It is a critical step, as the final depth of diffusion, or

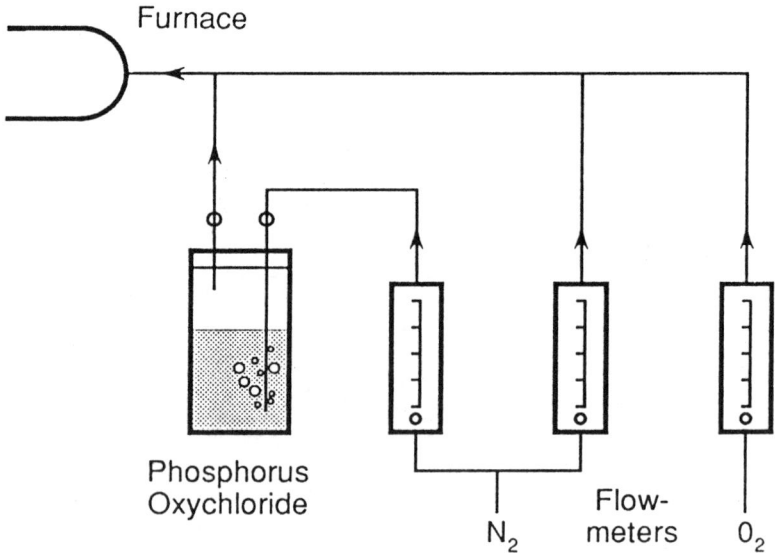

Figure 4.10 Liquid source diffusion method.

junction depth, as it is called, will be determined by it. In turn, junction depths will determine MOSFET and BJT device characteristics. It should be noted that *lateral diffusion, a* sideways diffusion under the masking oxide, also occurs. This effectively broadens the lateral dimensions of p-n junctions, which makes the fabrication of small devices more difficult.

Flying Ions

One of the unfortunate drawbacks of doping by diffusion is that precise control of dopant concentration and depth is difficult to achieve, particularly for very low dopant concentrations. A process called *ion implantation* can overcome this problem and help to reduce the lateral spread of dopant by "shooting in" accurate amounts of dopant atoms by the use of a *particle accelerator* known as an *ion implanter.* (Remember, an ion is effectively a charged atom and, as such, can be accelerated and deflected by electric and magnetic fields.)

The ion implanter (Fig. 4.11) consists of a *source* of dopant ions, which can be either a solid or a gas containing the appropriate dopant element; a *mass selection magnet* to ensure that only the desired dopant is chosen (other impurities are rejected); and an *accelerating column* which propels the ions toward the wafers by the use of high voltages, typically up to 200,000 volts. The *depth* of penetration of the ions into the substrate is controlled by the accelerating voltage. The quantity of the dose hitting the wafer is directly related to the beam current, which is merely the number of ions traveling down the column (regulated by a shutter). Both factors may be accurately controlled by the implantation equipment. Since the area of the ion beam tends to be small in comparison to the area of the wafer, the beam is scanned to cover the entire surface (or the wafer is moved to intercept the beam over its entire surface).

Ion implantation may be used to control critical MOSFET characteristics such as threshold voltage. This is the gate voltage at which the device begins to turn on, and is critically dependent on the amount of impurities in the channel between the source and drain. By implanting dopant through the gate oxide (before the gate electrode is deposited), the threshold voltage may be precisely adjusted to a desired level, so that the device is clearly off for a small gate voltage and on when a larger voltage is applied. An implant may also be used to turn an enhancement mode device to depletion

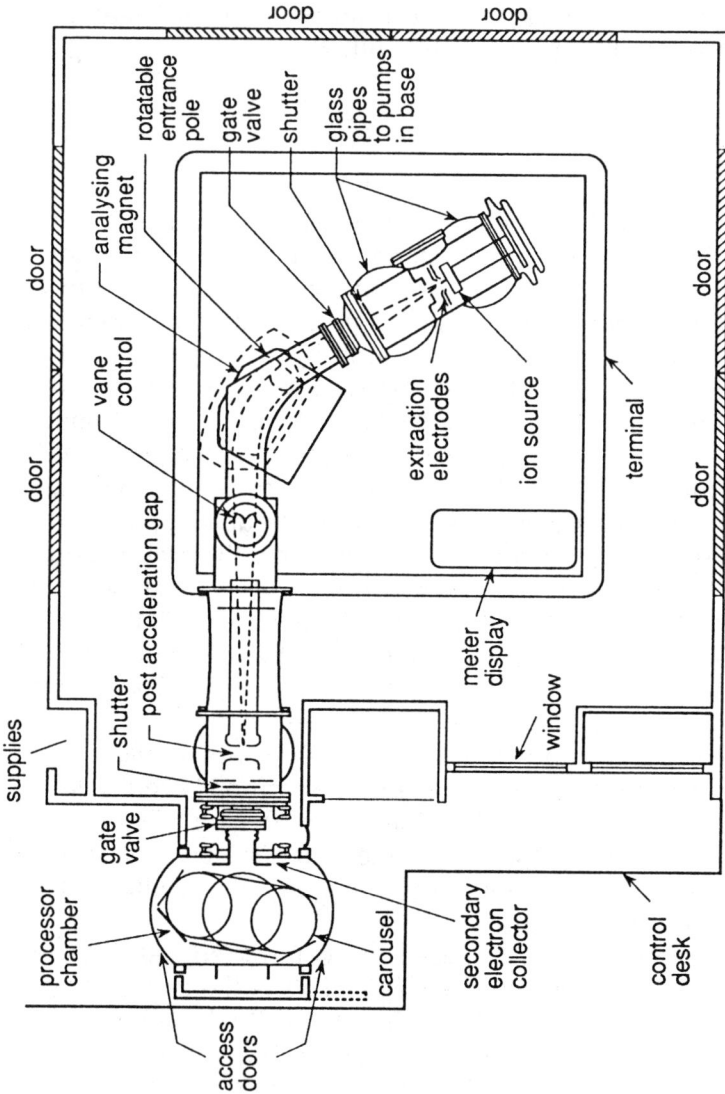

Figure 4.11 Ion-implanter schematic—source Lincott.

mode (see Chapter 6). Only ion implantation is accurate enough to allow this. The technique may also be used with higher doses to completely replace diffusion steps in MOSFETs and bipolar devices, especially in small geometry circuits where accurate control of dose is vital.

The main problem with implantation is that the ions "smash" the surface of the substrate and cause damage to the crystal. Some dopant remains electrically *inactive,* which means that they do not rest on appropriate positions in the semiconductor lattice and therefore cannot provide electrons or holes. To remove these problems, the semiconductor has to be *annealed.*

Heat Treatments

Annealing is the heat treatment of a semiconductor to remove damage caused by ion implantation or to activate dopant elements (cause them to move to lattice sites and become electrically active). The heating of metals on semiconductors to promote a reaction is sometimes called annealing but is more accurately called *sintering.* There are many cases where we want to promote reactions between metals and silicon. For example, to form a good electrical contact between aluminum and silicon, we must heat the two together so that they intermix slightly. To form a silicide, we must react a transition metal thermally with silicon.

Annealing is usually performed in a furnace with a *neutral* or *reducing* ambient. A neutral ambient, such as *nitrogen* gas, helps to prevent unwanted oxidation of the metals or other reactions during heat treatment, as the nitrogen will react with the metals only at extremely high temperatures. A reducing ambient, such as *hydrogen* or *forming gas,* is better in many cases as it can actually reverse oxidation processes. One of the drawbacks of annealing semiconductors for the purpose of activating dopants is that, since the process is performed at 600° C or greater, there can be appreciable unwanted movement of the implanted dopant during annealing by diffusion. This removes the advantage of being able to place the dopant accurately by ion implantation. Therefore, annealing times should be kept to a minimum to reduce redistribution of dopants.

Transient annealing uses high temperatures (>1000° C) and short times (<100 seconds) to maximize the dopant activation effect and repair implant damage and to minimize the redistribution of impurities. Methods used in transient annealing include high-energy electron and laser beams, hot graphite strips, and high-power halogen lamps, the last being most common in commercially available systems. All of these methods have a low *thermal inertia*; hence the wafers heat up and cool down rapidly. Short-term heating is also being used for other processes, such as oxidation and CVD, in order to produce thin layers in a highly controlled fashion.

A Typical Process

Figure 4.12 shows a typical process used to fabricate an IC. In this case, the technology is NMOS and the sequence shows only the creation of a single MOSFET. Of course, in a typical circuit, many devices would be fabricated simultaneously using this process. This particular technology was chosen because it provides many good examples of the processes used in the fabrication of many different types of IC. This section describes the process step by step.

1. A thin silicon dioxide layer, the *buffer* or *initial oxide*, is grown to protect the surface of the silicon, and a layer of silicon nitride is deposited by LPCVD on top. The silicon nitride is used to prevent oxidation in a later process step in the areas which will eventually contain the transistors. These device areas are called the *active areas*, as opposed to the *field areas*, the regions of the circuit which have most of the interconnect running in them. The use of nitride in this way is called the *local oxidation of silicon (LOCOS)* process.
2. The first photolithography step is carried out (using the "active area" or "field" mask) to delineate the two types of region in the circuit.
3. The nitride and oxide are etched using a CF_4 plasma, and the *field implant* is performed. The ions go into the substrate only where the nitride has been removed (the eventual field areas). The combined thicknesses of photoresist and/or (depending on the exact process used) nitride and oxide prevent penetration in the active areas. This implant helps to prevent the occurrence of parasitic (accidental) transistors in the field regions. There is usually a danger of parasitic device formation whenever there is a conductor on top of oxide over the semiconductor, even in the

field areas. Great care must be taken to prevent this (see also step 4). The resist is then stripped.

4. The *field oxide* is now grown. The field oxide is thicker than the gate oxide so that the field areas do not accidentally behave as MOSFETs. Factors affecting MOSFET performance are discussed in Chapter 6, but at this point, it should be understood that the dopant concentration (adjusted by ion implantation in the last step) and oxide thickness are critical factors in making (or not making) a MOSFET. The field oxide does not grow in the areas covered by nitride. However, the change between the active and field area oxide thicknesses is not abrupt, as the oxide tends to *encroach* slightly under the nitride. This creates an effect called *birds beak*, which is one of the factors which limits the smallness of devices. The nitride and buffer oxide may now be removed.

5. The high-quality *gate oxide* is now grown. After this, a further implant, called the *threshold adjustment* implant, is performed. The dopant ions pass through the gate oxide and into the semiconductor to fine-tune the device characteristics. The field oxide is thick enough to prevent penetration in these areas.

6. The *polycrystalline silicon* layer, which will eventually form the gate electrode, is deposited by LPCVD. It may be doped in situ or prior to etching by diffusion or ion implantation, or it may be doped later.

7. Photolithography is performed using the poly or gate mask, and the poly-Si is etched in a CF_4 plasma to define the gates and gate-level interconnects. The resist is stripped.

8. The source drain regions are now diffused, or more likely, implanted using the patterned poly-Si as a shield for the underlying channel so that the dopant goes into the substrate only at either side of the gate electrode (and not underneath it). The benefits of this *self-aligned process* are discussed in Chapter 6. If the poly-Si was not doped previously, it may be doped by this step.

9. A thick layer of oxide is deposited by CVD (APCVD, LTO, etc.) to provide support and isolation for the metal interconnection layer. This layer is frequently called by many names including *pyro*, *PVX* and *silox*.

10. Photolithography is performed using the contact mask to define *contact holes* through the deposited oxide layer so that the metal can make contact with the devices in order to interconnect them.

The oxide is etched by wet etching, plasma etching, or a combination of these techniques.

11. The metal (aluminum) layer is deposited by PVD and patterned by lithography using the metal mask. The metal will go into the contact holes to some extent to complete the connection between layers. Good metal step coverage is extremely important so that these connections are as low resistance and as reliable as possible.

Note that in this simple device scheme, three levels of interconnection are available to us: the metal, the poly-Si, and the source drain diffused/implanted regions.   Unfortunately, in a self-aligned MOSFET process, we cannot have the poly-Si interconnect cross the source drain level interconnect; otherwise, by definition, we will form a device! (Think about it: the source drain connections cannot go underneath the field oxide, so the poly-Si must cross them in an active area.) Also, unless we use silicides/polycides, it is not a good idea to use too many poly-Si interconnections, as these tend to have a much higher resistance to current flow than metal. Further layers of metal may be added, one on top of another, separated by an *intermetal dielectric* and interconnected by *via holes*. This is not as simple as it sounds, however, as obtaining good planarization to aid the step coverage of subsequent layers is frequently a problem.

Trends in Processing

To finish this rather lengthy but extremely important chapter, we will look briefly at some of the innovative trends which are emerging in semiconductor processing. We will not discuss alternative or novel device structures in this section. These will be covered in Chapters 5, 6, and 7.

One trend which has been with us for some time is an increased use of computers in semiconductor processing (computers making computers!).   Many pieces of equipment are now *automated*, using digital controllers to monitor functions and basically run the

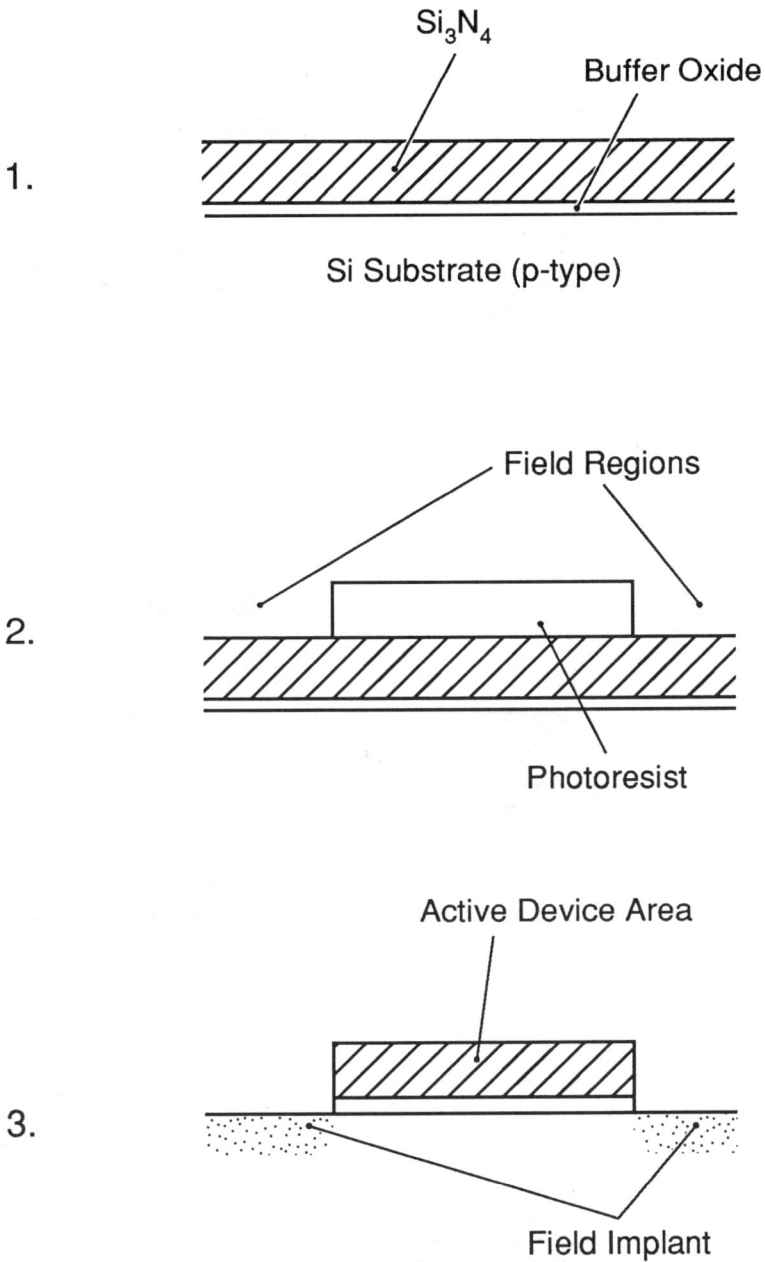

Figure 4.12 NMOS process outline (steps 1 - 11).

4.

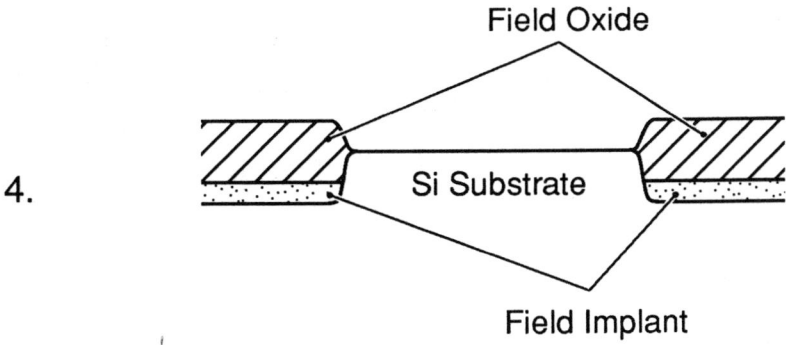

Field Oxide

Si Substrate

Field Implant

5.

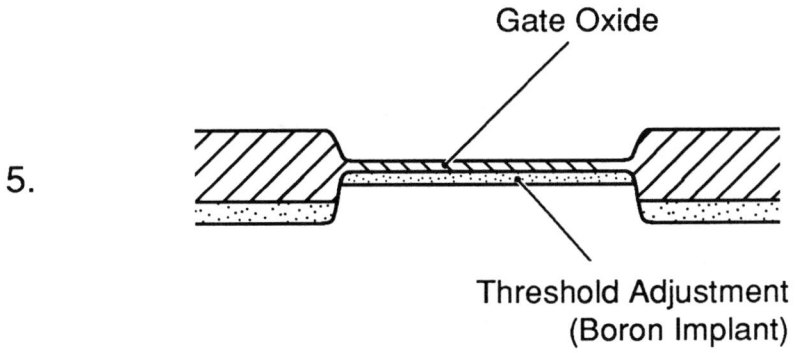

Gate Oxide

Threshold Adjustment
(Boron Implant)

6.

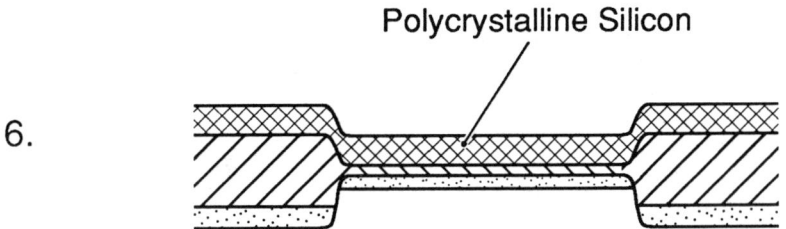

Polycrystalline Silicon

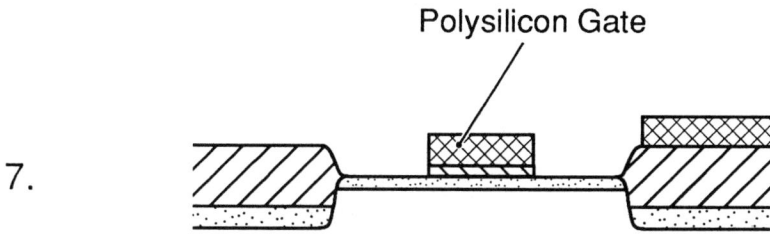

7.

Polysilicon Gate

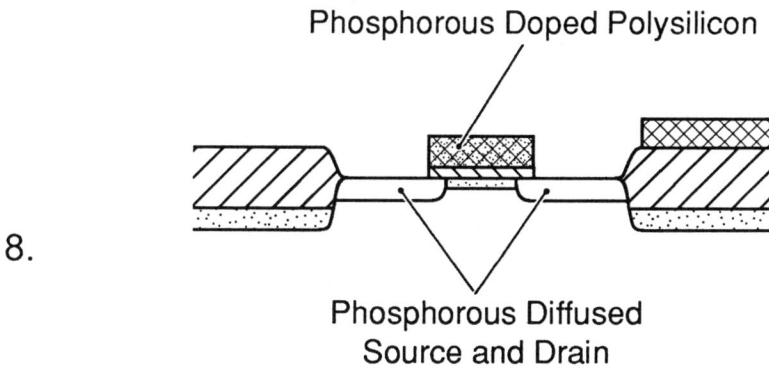

8.

Phosphorous Doped Polysilicon

Phosphorous Diffused
Source and Drain

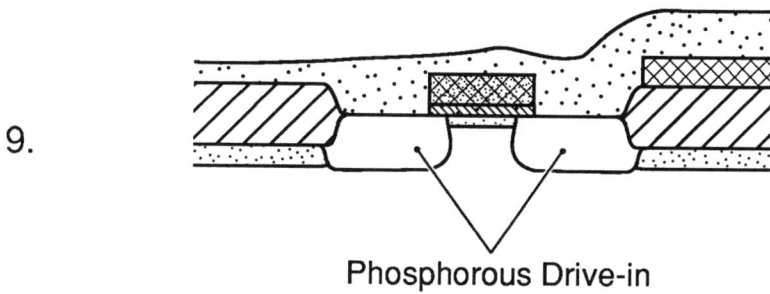

9.

Phosphorous Drive-in

Contact Windows

10.

Aluminum Contacts

N+   N+

11.

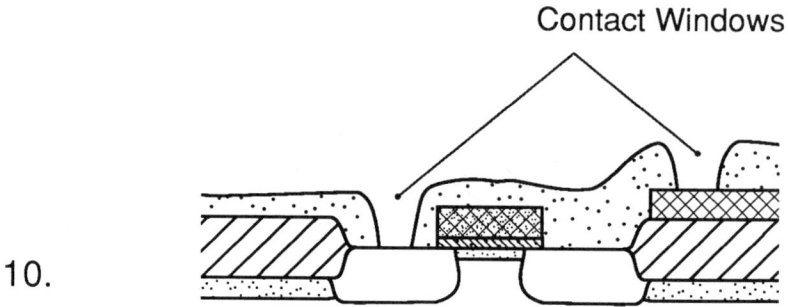

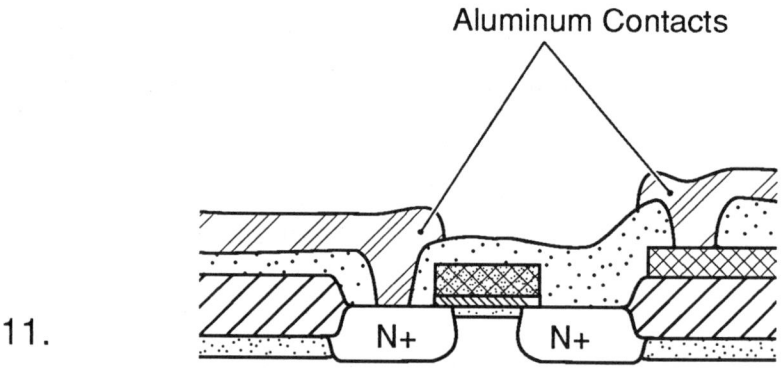

equipment more accurately than it could an operator in conjunction with a "dumb" analog controller. *Digital process controllers (DPCs)* also now have good communications skills in that they are able to "talk" to other systems by way of the *SECS II* information protocol. In addition, since it is often difficult and expensive to perform a series of experiments to determine the effects of a particular process or combination of process steps, computer programs are now available to perform *process simulation* with a reasonable degree of accuracy so that the experiments may be run on the computer. It is now becoming possible to link process simulation programs with design tools including *device* and *circuit simulators*. Eventually it should be possible to custom design and optimize processes to suit particular *application-specific integrated circuit (ASIC)* designs. With advanced software, the optimized manufacturing process design would proceed concurrently with product design. It is important to note that process and equipment automation encompasses both *information integration*, in which data are gathered and used intelligently throughout the entire process, and *mechanization*, the use of robots and other mechanical material transfer systems. Automation is often thought of as merely the latter, but the ability to utilize vast amounts of information about the product during its fabrication is much more important and will inevitably lead to better ICs.

Rapid thermal processing (RTP) is also becoming a commercially viable alternative to conventional long-term processing in standard furnaces. The low thermal inertia of the systems coupled with the high degree of control available with *single-wafer* approaches (only one wafer is processed at a time) makes this type of processing ideal for the accurate film growth and deposition. In many cases, it is possible to obtain information feedback directly from the wafer as it is being processed. This allows the control system to adjust process parameters in *real time*.

There are also potential alternatives to standard lithography and etch techniques. As mentioned previously, patterns may be written directly on resist using an electron-beam lithography system. This disposes of the need for a mask and therefore is ideal for simple circuits which must be produced with rapid *turnaround*. A good example of a circuit type which benefits greatly from this technology is *gate arrays*. These circuits have all the devices already processed and await the patterning of the final layer(s) of metal interconnect to wire the devices together to produce a circuit capable of performing a particular function. Electron-beam pattern definition allows fast *commitment* (connecting up) of the arrays. Other direct write techniques also show great promise. These include *focused ion beam (FIB)* for direct implantation of dopant ions and direct write deposition systems which use laser or ion beams to decompose vapors containing materials to be deposited. These materials are deposited only on the areas exposed by the beam; hence conducting tracks may be defined without the usual deposition-lithography-etch sequence. Of course, all of these serial (one area at a time) techniques are very slow in comparison to standard lithography and hence will probably tend to remain useful only for special circumstances (like gate arrays). However, it should be noted that FIB has attained widespread popularity in *mask repair* systems. Clear defects on the mask may be made opaque by hitting them with a localized high-dose ion beam. Opaque defects like extra pieces of chromium on masks are traditionally removed by high-power lasers which locally evaporate the extraneous material.

The combination of beam processing and precision deposition techniques such as molecular beam epitaxy has opened up the way to a relatively new process in the fabrication of structures and devices. *Nanotechnology* could be the ultimate destiny of microelectronics. In microelectronics, we currently have features within devices and circuits which have a minimum size of about 1 micron. In nanotechnology, we talk in terms of geometries which are tenths or even hundredths of a micron (a nanometer is one-thousandth of a micron). This is possible with electron-beam exposure and reactive ion etching of thin deposited layers. Although the technology is currently used to create structures used in research into the fundamental physics of devices, *nanoelectronics* could be the wave of the future, in which we will see *gigabit* (1 billion bits) storage capability on a single chip rather than the few million bit capacity we have today. Nanotechnology also makes possible novel electrical elements such as *quantum well devices*, in which the macroscopic behavior of electrons is determined by microscopic structures. Examples of these are discussed in Chapter 7.

It is interesting that many processing techniques are performed at low pressure or in a vacuum. This has the benefit that the room air cannot interfere with the process. It is possible that the processing plant of the future will be in the form of a single-vacuum tight chamber or a multiple connected chamber system rather than the type of facility discussed later in this text. We have already traveled some way down this road. *Cluster tool* equipment may now be bought in which several chambers, each performing a separate but related process, are connected by way of a central hub which uses an internal robot to transfer material between chambers.

To end in the realm of electronic materials, it is now becoming possible to think about the use of a new class of materials called *superconductors* in ICs. These superconductors have zero resistance to current flow and therefore should be the ideal interconnection material between devices in a circuit. However, at this time, no superconductors can operate at room temperature. The best of these, which have the unlikely name *high-temperature superconductors*, have to be supercooled by liquid nitrogen (which has a temperature just above -200° C). There are also problems in depositing and etching these materials. Much development work still has to be performed before we see superconductors in off-the-shelf ICs.

Summary

The fabrication of integrated semiconductor devices, circuits, and systems is an extremely complex process, undoubtedly one of the most complex in the industrial world. Fine line lithography and a variety of etching techniques are used to transfer the designed patterns to the materials which make up the devices. These materials may be deposited using physical or chemical vapor deposition methods or grown by reacting a gas with a substrate material. The p-n junction, the heart of most devices, is formed by diffusing or implanting an appropriate dopant, the latter process being used for modern ICs due to its inherent precision. The impressive range of processes is matched by the variety of materials used. These include semiconductors such as the silicon of the substrate, dielectrics such as the silicon dioxide used to insulate one layer from another, and conductors such as the aluminum used to connect the devices in a circuit together.

5

Transistors 1—
A Closer Look at Bipolar Devices

Bipolar Junction Transistors

In this chapter we will take a more detailed look at bipolar junction transistors (BJTs) and some related devices. In Chapter 2 we learned how these devices worked by first studying the properties of the p-n junction and then relating this information to the structure of the BJT. We also saw that BJTs can be of two types, depending on the *starting material* (substrate): *n-p-n* (n-type starting material and collector, p-type base, n-type emitter) and *p-n-p* (p-type starting material and collector, n-type base, p-type emitter). Figure 5.1 contains cross-sectional diagrams of a simple n-p-n transistor using planar technology. Remember that planar technology involves performing processing operations on one surface of a wafer; hence all the doped regions are put in from the front surface. In the early days of discrete transistors, the collector and emitter junctions were created on either side of a thin piece of semiconducting material.

Our starting material in Fig. 5.1 is n-type, which forms the collector of the transistor (a). We then create a large p-n junction by putting in a p-type area, first oxidizing the silicon and patterning the oxide to provide a mask which will withstand diffusion (or implant annealing) temperatures (b). This doped region will ultimately become our base. After this, a small n-type region is placed in the p-type area to create our second p-n junction, so that we now have a definite n-p-n structure (c). In real devices, to obtain a high degree of control over device characteristics, the emitter is frequently diffused using *heavily doped poly-Si* as the surface source of dopant (not shown). In this so-called *self-aligned emitter* scheme, the poly-Si remains to form the connection to the emitter. In our simpler case of Fig. 5.1, all is needed is to make electrical connections to the three regions from the top by way of a patterned aluminum wiring level (d). An oxide layer supports the metal wires, and contact holes provide a means for their connection to the substrate regions.

Unfortunately, there is a complication in real planar BJTs, as the collector region is lightly doped and hence has a relatively high resistance. This high *collector resistance* leads to poor devices and therefore has to be reduced. How this is done is quite cunning;. A highly doped n-type *buried layer* (also called the *subcollector*) is first put into a substrate, which is covered by an n-type *epi-layer* which will ultimately become the collector. Deep diffusions through the epi-layer connect the buried layer and hence the collector to the surface of the circuit. These diffusions typically surround the collector region to minimize the resistance. Figure 5.2 shows a cross-sectional diagram of a modern BJT, along with a micrograph taken on a *scanning electron microscope (SEM)*. The SEM can resolve very small features and hence is frequently used to view microelectronic components. The device in the micrograph has been treated using a special etch which allows us to see the position of the doped regions beneath the surface. The diffusion to the far left of the photograph is a p-type region which creates a p-n junction to *isolate* one BJT from another. This isolation normally has to surround each transistor (see later). Figure 5.3 shows a view of a typical BJT from the top, clearly showing the outer isolation region and connections to the collector (the inner ring), base, and emitter (the dots are *contacts* where the metal goes through the supporting oxide to connect with the regions underneath). In Figs. 5.2 and 5.3, the width of the metal lines is approximately 1.5 microns.

Although device models and the associated physics and mathematics are beyond the scope of this book, it is still useful to take a brief look at the electrical characteristics of a typical n-p-n BJT so that we may compare their performance with that of other major device types. Figure 5.4 shows how the current through the collector-emitter circuit (ICE) varies with the voltage across the device (VCE) for different currents flowing into the base (IBE). (Remember

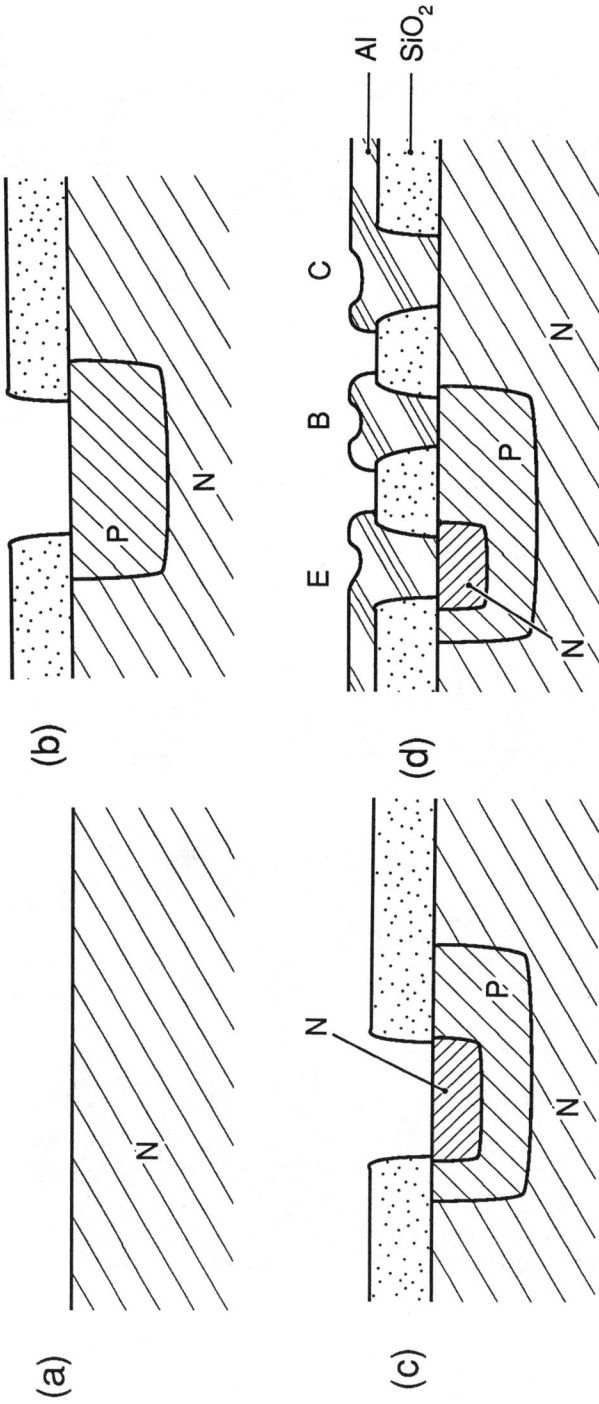

Figure 5.1 Formation of an n-p-n bipolar transistor.

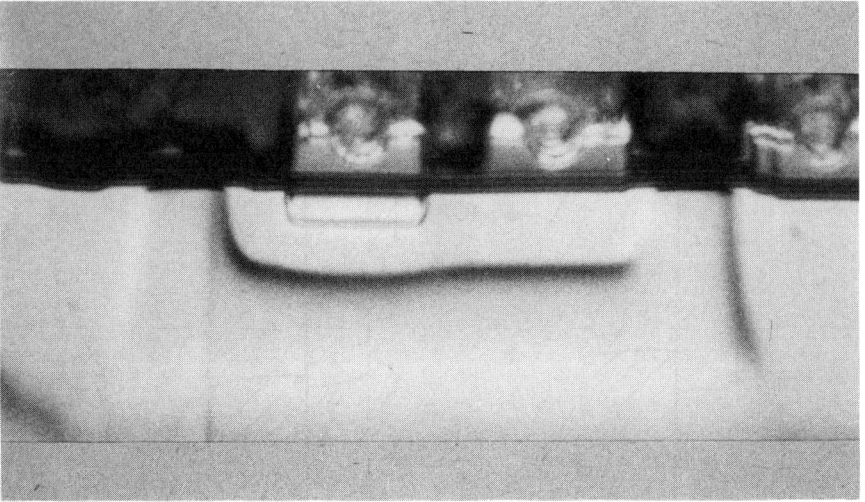

Figure 5.2 Micrograph of a BJT (cross section)—source CMD.

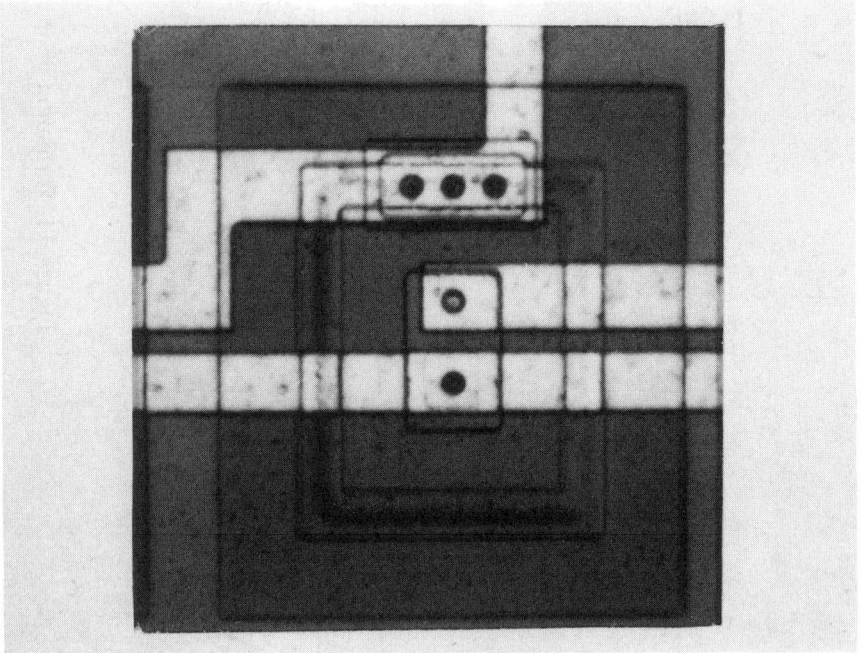

Figure 5.3 Micrograph of a BJT (top view)—source CMD.

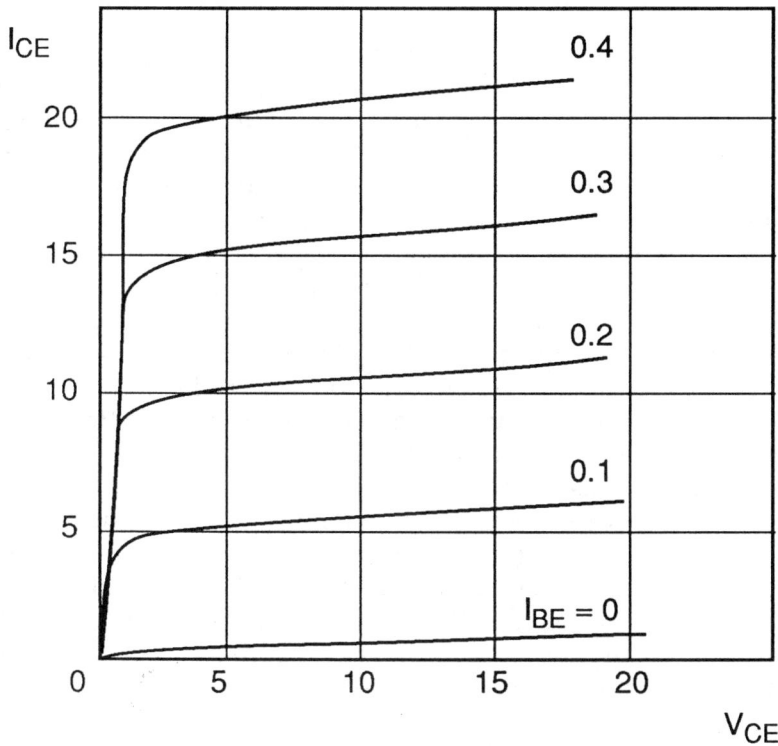

Figure 5.4 Electrical characteristics of a BJT.

that we have to apply a voltage to a circuit to get a current to flow; an increasing base voltage means an increasing base current.) When no current flows into the base (IBE = 0), which is essentially the same as no voltage applied to the base (see Chapter 2), the device is off and little or no current flows in the collector-emitter circuit. As we increase IBE (which is done by increasing the voltage on the base), the device begins to conduct; hence a current flows through it when we apply a voltage. The curves are not perfectly flat, as the internal resistances and junction potentials make the characteristics less than ideal. However, the device performs pretty much as expected and, in this case, has a *gain* of about 50 (the output current, divided by the input current or ICE divided by IBE).

In many applications which require silicon BJTs, n-p-n transistors are favored, as they can operate faster than similarly sized p-n-p devices. This is because the main current carriers through the device are electrons, which are more mobile in silicon than holes.

Heterojunction Bipolar Transistors

The *heterojunction bipolar transistor (HBT)* is a compound semiconductor cousin of the BJT. It is a bipolar transistor made from an n-type GaAs collector, a p-type GaAs base, and an aluminum gallium arsenide (AlGaAs) emitter. AlGaAs is a compound semiconductor of all three elements which forms the necessary emitter base junction. It is called a *heterojunction transistor* because heteroepitaxy is used to put the AlGaAs layer on the GaAs.

The advantages of this structure are improved frequency response due to low base resistance and high electron mobility, and wider temperature range of operation (-270° C to 350° C) due to the large bandgap of the compound semiconductors. Note that silicon, which has a smaller bandgap, becomes a rather poor semiconductor at high-temperature; therefore, we have to ensure that silicon-based devices are kept cool (125° C is a typical operational maximum). These factors make HBT devices very suitable for high-frequency applications in communications and signal processing, particularly in severe environments such as those of spacecrafts and cars. Unfortunately, they are not as easily integrated into complex circuits and may not be as readily mass produced as their silicon counterparts. They are therefore considerably more expensive.

Small Signal and Power Transistors

Bipolar devices have traditionally fallen into two main categories, *small signal* and *power devices*. Small-signal devices are used in applications in which information, in either analog or digital form, is processed (for example, in an audio preamplifier or a large mainframe computer). Here we are dealing with relatively small voltages and currents. Power devices are used to control large currents and voltages (for example, a power amplifier which drives loudspeakers or a dc electric motor control). There are a number of special device types which could also fall into these categories, but we will focus on BJTs here and discuss the others in Chapter 7.

For BJTs, the main differences between small-signal and power devices are *size* and *packaging*. Small-signal transistors have to be physically small, with a thin base region. The base has to be thin so that its resistance to current flow is low. This is very important for circuits which deal with small currents and voltages. As an added benefit, if the base is thin, carriers can move across it in a relatively short time; hence small-signal transistors can also be fast. Of course, we could also get the carriers themselves to move faster by using compound semiconductors; that is why HBTs can have such high operating speeds. Small-signal silicon devices are relatively easy to integrate into large, complex circuits due to the fact that they are compact and carry small currents. In any case, the packaging (a subject dealt with in more detail later in this text) tends to be relatively simple for small-signal devices. Whenever current flows in a material which has resistance, the material heats up. The amount of heating is proportional to the resistance and to the square of the current. Single transistors or small numbers of devices in integrated form do not carry a great deal of current in total and therefore do not heat up very much. They may therefore be packaged inexpensively in plastic. However, when large numbers of transistors are integrated into a single circuit, even though the current flow through each of them is small, the sum of the current flow through all of them is large. Therefore, in packaging them, we may have to treat the circuit in much the same way as a power device.

Power devices have to be physically large to take large currents. In a conductor of electricity, a large cross section means that the resistance per unit length is low. If the devices have to take high voltages, they also have to be large; otherwise, they can breakdown. Semiconductor junctions can withstand only relatively low voltages. In a breakdown condition, a p-n junction can no longer restrict current flow in the reverse bias condition; hence the device does not function. In a high-voltage BJT, the base has to be relatively thick so that it can withstand the high voltage across the device without breaking down. With large currents or voltages, the heating effect can be quite severe. Power devices are therefore frequently put in metal packages with *heat sinks* attached to the package to help dissipate heat during operation. The heat sink is either a metal tab which allows the device to be bolted to a larger metal surface to conduct the heat away or a finned metal attachment which radiates heat. If these are not used, the heating effect may cause a condition known as *thermal runaway*, which destroys the devices.

Comparing Bipolar and MOS Technologies

We saw in the first section of this chapter that the structure of the BJT was fairly simple in theory but quite complex in practice. Even though the part which controls the current is small, due to the need for isolation structures and large collector connections around the devices, real-life integrated BJTs are large. More to the point, they are considerably larger than their MOSFET counterparts, which do not require the same degree of isolation and subsurface connection. Therefore BJT technology has a poorer *packing density* potential than MOS technology. The circuits with the largest number of transistors, such as microprocessors and ULSI, memories are therefore MOSFET based.

A further problem with BJT devices is that they draw current through the base circuit. In MOS devices, only a tiny *leakage* current flows into the gate. We therefore say that a BJT has a lower *input impedance* than a MOSFET. This is an important factor in applications which require high input impedance, such as the amplification of very small biological signals. In this case, a MOSFET input may be put on a BJT circuit. BJTs also tend to be poorer switches than MOSFETs, allowing more current to flow in the off state than a comparable MOS transistor. Thus BJT circuits tend to draw more current and hence *dissipate* more power (heat up more). This is a further limitation to packing density, because if we attempt to put too many BJTs together without efficient cooling, the heat generated would destroy the circuit.

Despite their packing density drawbacks, BJTs still have many advantages over MOS devices and therefore they still have an important place in IC technology. The MOS structure is effectively a capacitor, and, as we saw earlier in this book, capacitance can slow electrical signals. BJTs do not have this problem, and if the base is thin, they can operate at very high speeds. They are therefore the ideal choice for IC sin large, high-speed mainframe computers. In addition, it is easier to make high-power devices using BJT fabrication technology. Many MOSFET circuits today have BJTs integrated into their outputs to enable them to supply larger currents to the external circuit they are connected to without the need for external discrete power components.

Summary

BJTs are an extremely important circuit elements in electronics. They are formed by creating back-to-back p-n junctions in a semiconductor substrate. They may have a p-n-p or n-p-n structure, depending on the starting material and the desired use, but n-p-n transistors are favored in many silicon BJT applications, as they operate faster. HBTs, based on compound semiconductors, are faster still and are used for high-frequency analog applications. A small current or voltage on the input (base) circuit controls a larger current in the output (collector-emitter) circuit. Power BJTs are physically large to enable them to withstand high currents and voltages, and small-signal devices are small to allow them to operate quickly. Circuits made with BJT devices tend to consume more power and are physically larger than their MOS counterparts, but they are typically faster.

6

Transistors 2— A Closer Look at Field Effect Devices

Junction and Metal-Semiconductor Field Effect Transistors

The *junction field effect transistor (JFET)* is one type of *field effect transistor (FET)*. The main steps in the making of a typical *n-channel depletion-mode JFET* structure using planar technology are shown in Fig. 6.1. In this case, a p-type substrate (a) is used, and the n-type channel region is diffused into it (b). A p-type gate diffusion is then put in, so that only a narrow subsurface n-channel region remains (c). A narrow channel allows lower gate voltages to be used to control the current flow from source to drain. We then put on the metal-supporting dielectric and the metal connections to the source, drain, and gate as before (d).

The electrical characteristics of this device are shown in Fig. 6.2. Recalling our discussion in Chapter 2, as we increase the voltage on the gate in the negative direction, we reverse-bias the gate junction and form a depletion region in the channel. This depletion region becomes wider as the magnitude of the applied voltage increases and acts to reduce the current flow from source to drain. We can clearly see this effect in Fig. 6.2, where the device conducts (is on) when no

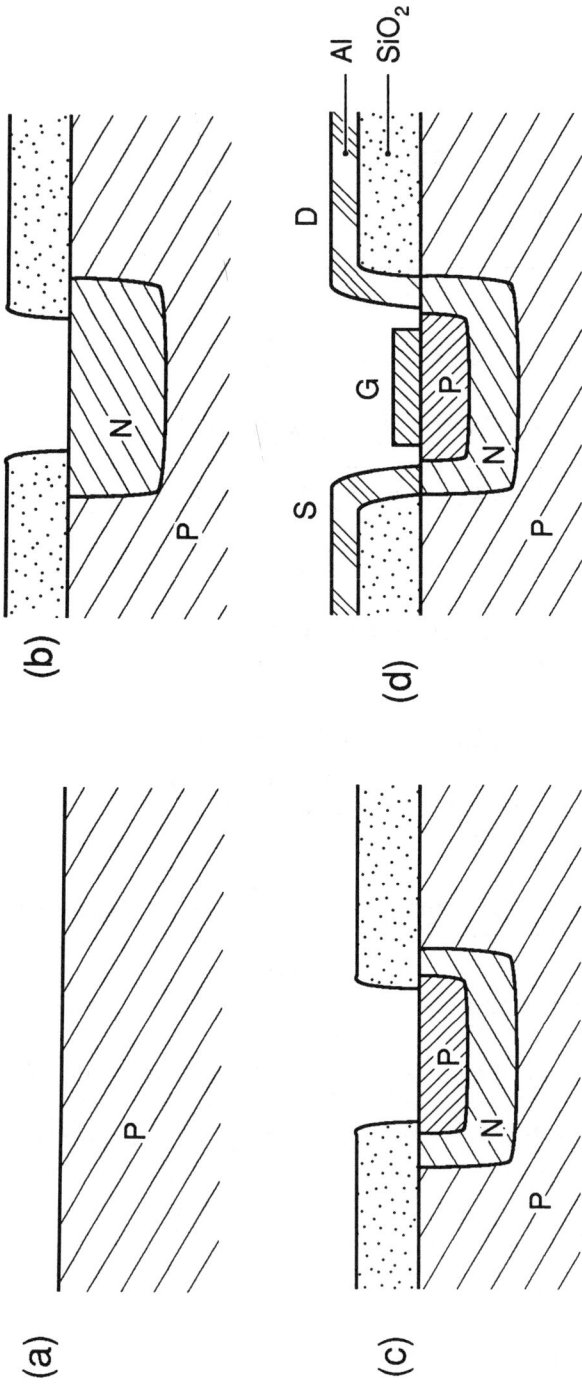

Figure 6.1 Formation of an n-channel depletion-mode JFET.

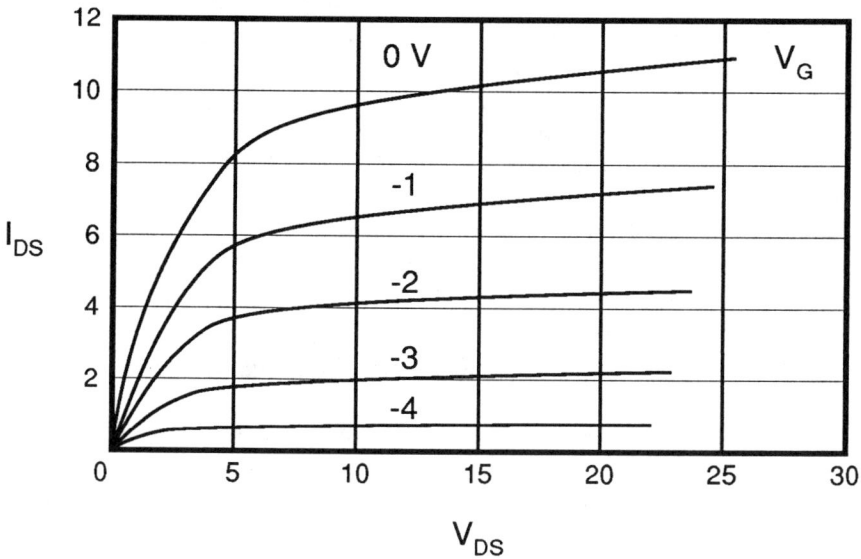

Figure 6.2 Electrical characteristics of a JFET.

voltage is applied but conducts less for larger negative gate voltages. This is why we call this device a *depletion-mode FET*, as it is normally on and we have to deplete the channel to turn it off. Note that in the p-channel, the starting material is n-type and the polarities are reversed.

The *input impedance*, the resistance to current flow in the gate circuit, of the JFET, is not as high as that of the MOSFET due to *junction leakage* (the small but nevertheless measurable trickle of current across a reverse-biased p-n junction). However, the input impedance is higher than that of the BJT which requires some current flow in the base circuit to make it work at all. Also, it is a *normally on* device, which makes it nonideal for many circuits, including most forms of digital logic. Therefore, this device is not frequently used in modern ICs, except in the input stages of special analog circuits, which consist primarily of BJTs, called *operational amplifiers* (op-amps). JFETs are ideal for this application, as they can easily be fabricated along with BJTs and provide a better input impedance for the op-amp, making it more sensitive to small input signals.

Another type of FET is the *metal-semiconductor field effect transistor (MESFET)*. The operation of the MESFET is essentially identical to that of the JFET, but a *metal-semiconductor rectifying contact* is used instead of a p-n junction for the gate of the device. This is one of the curious things about semiconductors: if the metal-semiconductor junction is heat treated, the junction resistance generally will be low and current will flow through it in the normal fashion. However, if the junction is not heat treated, its resistance will be high. If, in addition, the "electrical" difference, called *barrier height*, between the gate metal and the semiconductor is large enough, the abrupt metal-semiconductor junction acts like a p-n junction. Therefore, to form a rectifying contact, we merely have to chose an appropriate metal for the semiconductor we are using. In an actual device, a different metal is used for the source-drain connections or the same metal is heat treated to allow them to act like *ohmic* (nonrectifying) contacts.

A typical MESFET structure is shown in Fig. 6.3. Note that the device is based on an insulating or semi-insulating (very high resistance) substrate with a conducting epitaxial layer on top to confine the current flow to the area near the surface, where it may be controlled by the gate. Once the epi substrate is formed (which is the tricky part for compound semiconductors), MESFETs are simple to make. The characteristics of very small MESFETs are much more easy to control by fabrication than those of small JFETs, as we do not have to diffuse a gate junction precisely. Small MESFETs are very fast because the distance between source and the drain can be made small. They also lack the gate capacitance of similarly sized MOSFET devices. MESFETs are almost always based on compound semiconductors, so that the advantage of the higher carrier mobility in these materials may be used to maximize device speed. They therefore tend to be used for ultra-high-frequency applications (microwave communications), but mainly in discrete form. However, as with JFET, the input impedance is not as high as that of the MOSFET, and the on and off resistances are not particularly good (they are not very good switches).

Metal-oxide-semiconductor Field Effect Transistors

The *metal-oxide-semiconductor field effect transistor (MOSFET)*, also called the metal-insulator-semiconductor FET (MISFET) and the *insulated gate field effect transistor (IGFET)*, differs from the JFET and MESFET in that the gate is insulated from the substrate by a thin layer of dielectric

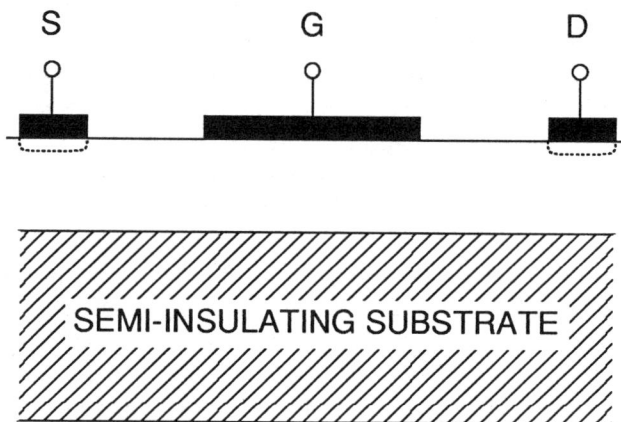

Figure 6.3 Structure of a MESFET.

material, usually an oxide of silicon. This unfortunately creates what is known as a *metal-insulator-semiconductor (MIS) capacitor* as part of the transistor. This potentially makes it the slowest of the devices we have examined so far, as capacitance in any electrical circuit has the effect of slowing its operation. However, it has many advantages over the other device types. Most important, it has the highest input impedance, due to the fact that the gate is isolated from any current flow path by the gate insulator; it may be made small; and it generally does not consume large amounts of power and therefore does not suffer from heat problems. These factors, along with the fact that it is easy to fabricate MOSFETs close to one another on the same substrate, means that many MOSFETs may be packed together in a very small area. Whereas the BJT is favored in very-high-speed computer circuits, the MOSFET is the obvious choice for complex microprocessors and high-capacity integrated memories.

There are two types of MOSFET, depending on the starting material and the fabrication sequence used: *n-channel* or NMOS and *p-channel* or PMOS. There are also two forms of the n- and p-channel MOSFET, which have very different electrical characteristics: the *enhancement-mode MOSFET* and the *depletion-mode MOSFET*. The enhancement-mode device requires a gate voltage to switch it on, whereas the depletion-mode device requires an opposite gate voltage to turn it off. For instance, in enhancement-mode NMOS, a positive gate voltage is required before the device will allow current to flow in the source-drain circuit. In depletion-mode NMOS, the device allows some current to flow with no applied gate voltage, and actually requires a negative gate voltage to turn it off. The polarities are reversed for enhancement- and depletion-mode PMOS.

We have already looked at a typical fabrication sequence for an actual NMOS device in Chapter 4, which provided excellent examples of some of the more common processing tricks used for forming ICs. That particular sequence was for a modern MOSFET structure, which we will examine a little later in this section. However, in keeping with the theme of this chapter, we will now look at a more basic MOSFET fabrication sequence so that we may make some intelligent comments regarding its benefits and drawbacks. The sequence for a typical n-channel enhancement-mode MOSFET structure is shown in Fig. 6.4, and its electrical characteristics are presented in Fig. 6.5. Note that in the p-channel case, the starting material is n-type and the polarities are opposite. As we can see from Fig. 6.4, we begin by diffusing the n-type source-drain regions into the p-type substrate (a). A gate oxide is then formed by oxidizing the silicon substrate (b). Then the gate metal is deposited on top of this oxide and patterned (c). This same metal layer is also used for the connections to the source and drain regions. A thick field oxide layer supports the metal wires, and contact holes provide a means for their connection to the substrate regions below (d).

Although this is the most simple way of making a MOSFET, it is not the best way. As mentioned before, the capacitance between the metal gate and the substrate will act to delay electrical signals. This cannot be avoided, as it is an integral part of the device. However, *parasitic capacitances* are also formed when the gate metal overlaps the source and drain regions, and in many respects these are highly undesirable. We will see how these capacitances may be reduced in the next section.

Turning now to the electrical characteristics shown in Fig. 6.5, we can see that this NMOS transistor behaves much as predicted in Chapter 2. With no applied gate voltage (Vgs = 0), practically no current flows for any voltage across the source-drain circuit. However, as we increase the voltage on the gate in the positive direction, nothing happens at first until we exceed a particular gate voltage called the *threshold voltage* (VT). It is at this point that we begin to see a current flowing in the source-drain circuit when we apply a sufficiently large voltage across it. Figure 6.6 shows this

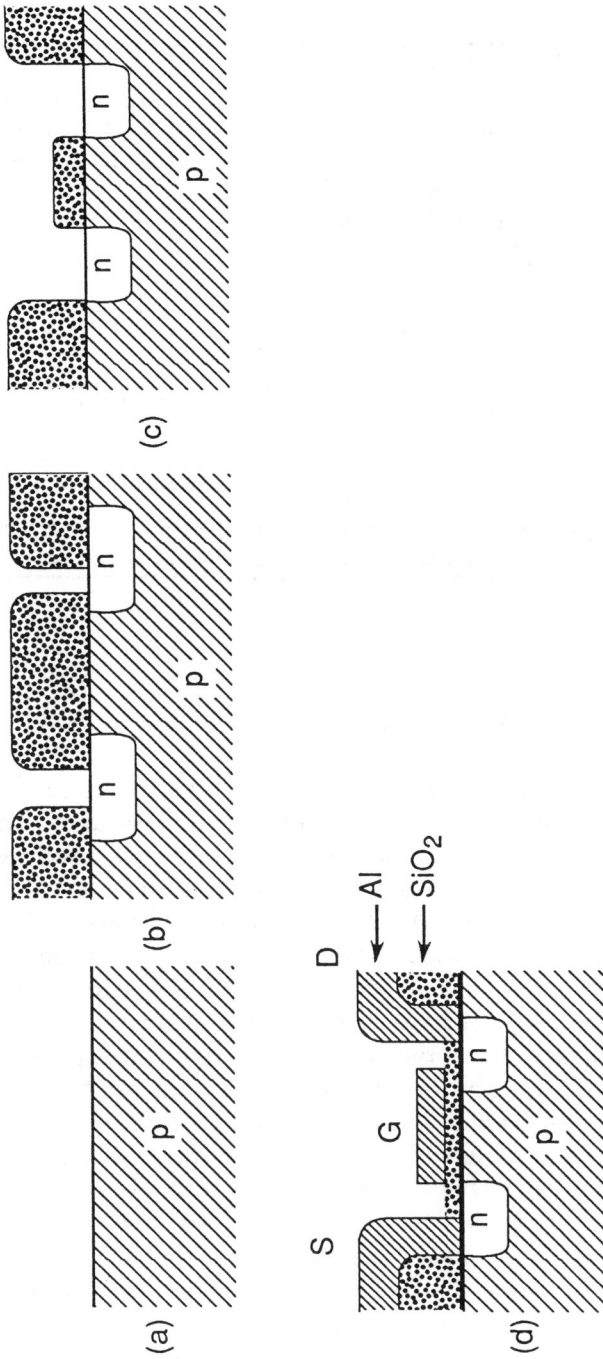

Figure 6.4 Formation of an n-channel enhancement-mode MOSFET.

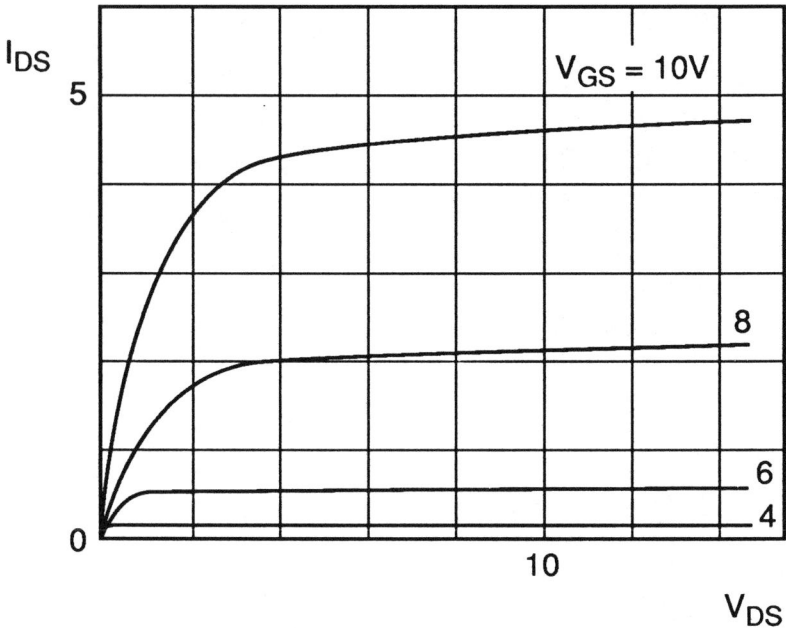

Figure 6.5 Electrical characteristics of a MOSFET.

threshold effect by plotting the current flowing through the device against gate voltage for a fixed source-drain voltage. This effect is due to the fact that some of the gate voltage is used to *invert* the semiconductor in the first place in order to form a conducting channel between the source and the drain. The actual value of the threshold voltage depends on the amount of dopant in the semiconductor, the thickness of the gate oxide, the amount of charge associated with the oxide (oxides tend to have charge fixed in them), and the material used for the gate. VT is typically around one volt in practical devices. The threshold voltage is typically fine-tuned by *implanting* dopant through the gate oxide prior to metal deposition so that a precise value may be obtained (this is important for the operation of many types of circuits). This is one of the factors which make MOSFETs ideal for digital circuits, as they are good switches with a definite off state and a good on state. Also, since the gate is insulated from the rest of the device, virtually no current will flow into the gate. The input impedance is therefore extremely high, which makes this type

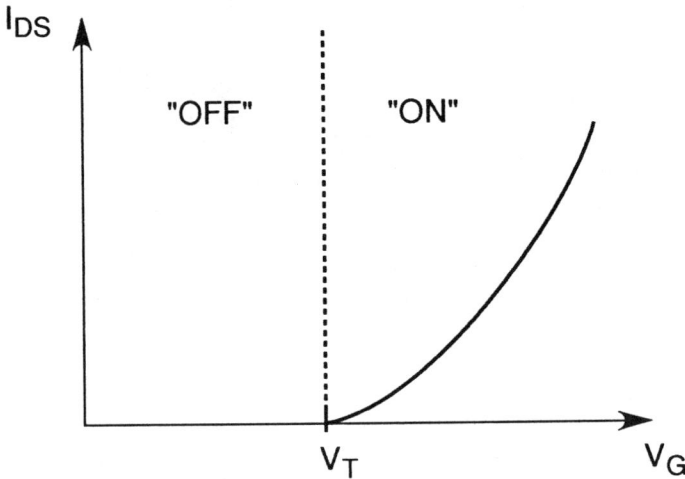

Figure 6.6 The threshold voltage effect in MOSFET.

of device ideal for extremely low-input current applications in analog circuits.

PMOS was developed before NMOS due to a better ability to control the threshold voltage with oxides grown on n-type substrates. NMOS is preferred in modern circuits, as the *mobility* (speed) of electrons, the current carriers in NMOS, is greater than that of holes, the current carriers in PMOS, by a factor of 3. The first MOSFET-based microprocessors therefore consisted of NMOS devices.

A typical depletion-mode NMOS structure is shown in Fig. 6.7, with its electrical characteristics in Fig. 6.8. The device is formed in much the same fashion as the enhancement-mode device, except that a *channel adjust* implant is put in after the gate oxide is grown. The surface of the semiconductor is deliberately made slightly n-type by this implant, so that there is always a conducting channel between the n-type source and drain regions. To close this channel off, we have to invert it (to p-type) by applying a negative gate voltage; hence the threshold voltage is actually negative. The net effect is a device which is on with zero applied voltage. Note that for the p-channel case, the starting material is n-type, the channel adjust is p-type, and the polarities are opposite. A positive voltage allows greater source-drain conduction, but a negative applied voltage is required to turn the device off (for the n-channel case). This device is used almost

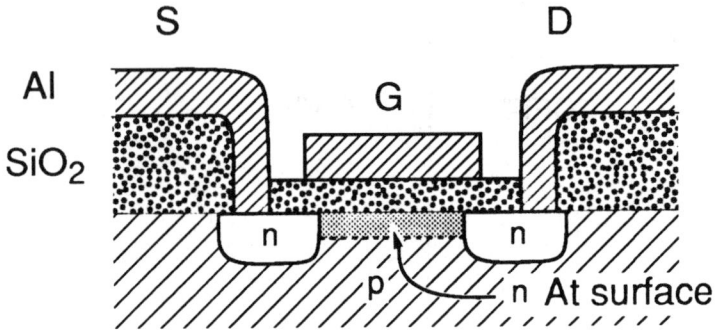

Figure 6.7 Depletion-mode MOSFET.

exclusively as a load in digital circuits (this concept is discussed in Chapter 8).

There are many applications for which it is advantageous to combine NMOS and PMOS devices in the same IC. When we

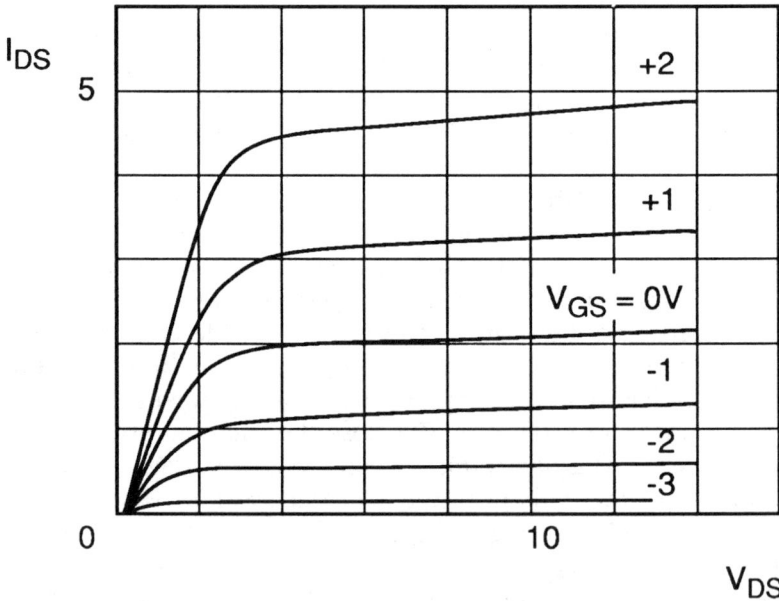

Figure 6.8 Electrical characteristics of a depletion-mode MOSFET.

combine both device types, the result is called *complementary MOS* or CMOS. In the simplest CMOS scheme, we begin with a substrate of one type, for instance n-type, and form the appropriate devices in it, (for our n-type substrate, these would be p-channel). The opposite type devices, then-channel FETs, are formed in a *well* of opposite type material, p-type in our example, formed in the substrate. If the starting material of the substrate is n-type, the well is p-type; this is called *p-well CMOS*. A typical p-well CMOS structure is shown in Fig. 6.9. Note that in the n-well CMOS, the starting material is p-type. An alternative scheme is also used in which a lightly doped epi is the starting material, and both p- and n-wells are put into it. This *twin-tub* approach allows a high degree of control over the doping of both wells, and hence good control over n- and p-channel device characteristics.

CMOS technology has the high-input impedance characteristics of MOS but may be made to operate with extremely low power consumption (this is discussed in more detail in Chapter 8). CMOS is best used for applications which require low power consumption (e.g., battery-powered circuits or very dense circuits which would overheat if power consumption was high). Unfortunately, CMOS has a curious drawback; an examination of Fig. 6.9 reveals that placing a well of p-type material in an n-type substrate, and diffusing n- and p-type areas into the well and substrate, respectively, creates structures which are reminiscent of the BJT. In fact, we create parasitic vertical n-p-n (drain-well-substrate) and horizontal p-n-p (drain-substrate-well) transistors by making CMOS. In CMOS, these incidental devices are naturally "wired" in such a configuration that particular conditions, such as the output accidentally being connected to the positive supply voltage, will create *latch-up*. In this condition, both transistors keep each other in the on state, even when the initial stimulus is removed. A high current can flow through the substrate, possibly destroying the devices in severe cases. We can significantly reduce this effect by putting in *guard rings* to separate the devices. These are more compact than in the bipolar case, consisting merely of an n-type diffusion next to the p-type drain and a p-type diffusion next to the n-type drain (the guard rings are put in during the source-drain diffusions). For very dense CMOS, *trench isolation* may be used instead. In this case, a narrow trench is etched between the n-channel and p-channel devices and then filled with a dielectric.

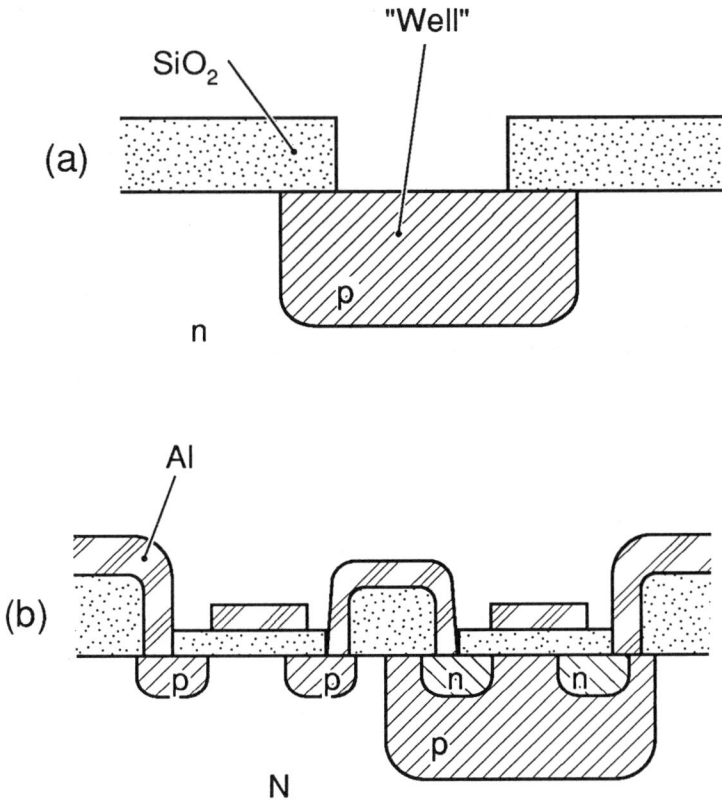

Figure 6.9 Formation of CMOS. (a) Formation of well. (b) Finished devices.

Gate Electrode Materials

In many respects, the most important element in a FET is the gate. The gate material, gate insulator, and *insulator/semiconductor/interface* are extremely important to device and circuit operation, as, along with the quantity of dopant in the substrate, they determine the threshold voltage of the MOSFET. In addition, the gate electrode may be extended beyond the transistor to act as an interconnect between devices in the circuit. Therefore, as well as being compatible with the MOSFET itself, it must also be a good conductor.

A variety of gate materials are used for MOSFETs. They all have certain advantages and disadvantages, which we shall list in this section.

Aluminum. This material forms the gate in so-called *metal gate* PMOS, NMOS, and CMOS. Aluminum is an excellent conductor, is easily deposited by evaporation or sputtering, and may be readily patterned by lithography and etching. Metal gate MOS is also the simplest MOS technology and therefore the least expensive to manufacture. Unfortunately, the use of this metal for the gate affects the voltage at which the device turns on due to an electrical property called the *work function difference* between the metal and the silicon (it adds a component to the threshold voltage). This must be compensated for, particularly in CMOS, where the threshold voltages of the PMOS and NMOS devices must be balanced. Worse still, aluminum is not a *refractory* metal, that is, it cannot withstand high temperatures. Subsequent processing is therefore limited. For this reason, since diffusion and implant annealing are high-temperature processes, the diffused or implanted source-drain regions have to be formed before the gate metal is added. In using this approach, One must ensure that the gate covers all of the channel. If a gap is left where the gate does not cover the channel, that part of the channel will not be inverted and the device will not function as a transistor. This is more difficult than it seems, as there are frequently dimensional shifts during processing due to misalignment, over- or under-exposure during lithography, and over-etching. With this in mind, to compensate for these shifts and ensure that the gate totally covers the channel, a gate overlap of the source-drain regions is necessary. Unfortunately, this creates a capacitance between the gate and the source and between the gate and the drain. These gate-source, gate-drain capacitors, as shown in Fig. 6.10, seriously affect the performance of the device and limit its speed of operation. This problem is particularly severe when device geometries are small, as the relative size of the overlap capacitance becomes larger. Therefore, metal gate MOS is generally found only in SSI, MSI, and some (>5 microns) LSI circuits.

Polycrystalline silicon. This material, which is also called *polysilicon* or *poly-Si*, is used in *silicon gate* MOS. It is a form of silicon which is easily deposited by low-pressure chemical vapor deposition and readily etched to form small structures, but it is a poor conductor

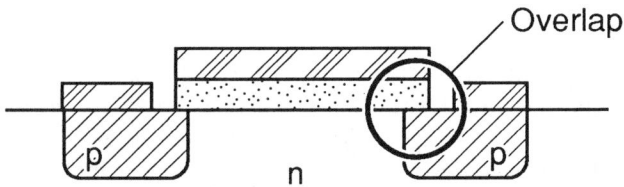

Figure 6.10   Example of gate overlap.

even when heavily doped.   However, it provides a minimal work function difference component to the threshold voltage and, more important, it is a good high-temperature material.   The source-drain regions may therefore be put in by diffusion or ion implantation with the gate in place.   Therefore no deliberate gate overlap is necessary. Since the gate acts as a mask during the diffusion or implantation of the source-drain regions, protecting the underlying channel from the dopant, the source and drain are naturally aligned with the gate, and the gate thereby covers the entire channel (Fig. 6.11).   This is called the *self-aligned* gate process.   Multiple levels of poly-Si may be used in devices to provide extra levels of interconnect, elements for large capacitors in analog circuits, or parts of special memory structures (see Chapter 7).   Silicon gate MOS is faster than its metal gate counterpart due to the small gate-source and gate-drain capacitances.   It also allows smaller devices to be fabricated, since we do not have to include an overlap factor in the design (and poly-Si is a bit easier to etch than aluminum), and it is therefore found in LSI and VLSI circuits.   Figure 6.12 shows a cross-sectional micrograph taken by scanning electron microscope of a silicon gate n-channel device in a p-well.   Once again, the substrate has been etched to show the position of the different type areas.   The top view of the device is shown in Fig. 6.13.   The width of the metal lines is about 1.5 microns in these micrographs.

Silicides.   This group includes the commonly used gate silicides *tungsten silicide* (WSi_2), *titanium silicide* ($TiSi_2$), *tantalum silicide* ($TaSi_2$), and *molybdenum silicide* ($MoSi_2$).   The silicides used in microelectronics are mainly compounds of transition metals (a group of materials which occupy the middle of the periodic table) and silicon, and have a good tolerance to high-temperature processing but are much better conductors than doped poly-Si. The interconnection lengths at gate level may therefore be longer and the conductors may be made

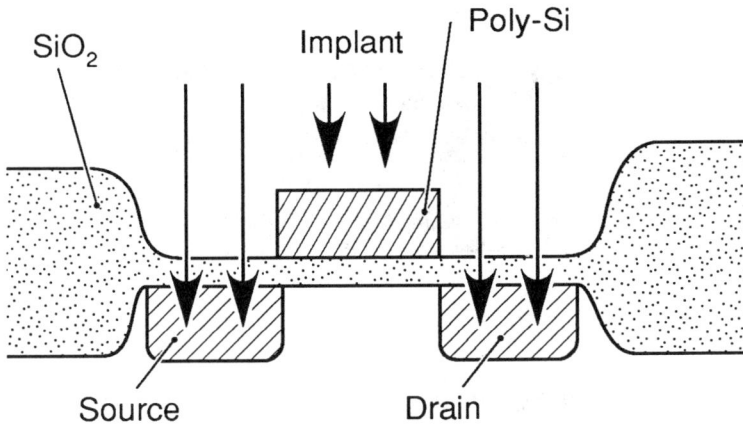

Figure 6.11 Self-aligned gate process.

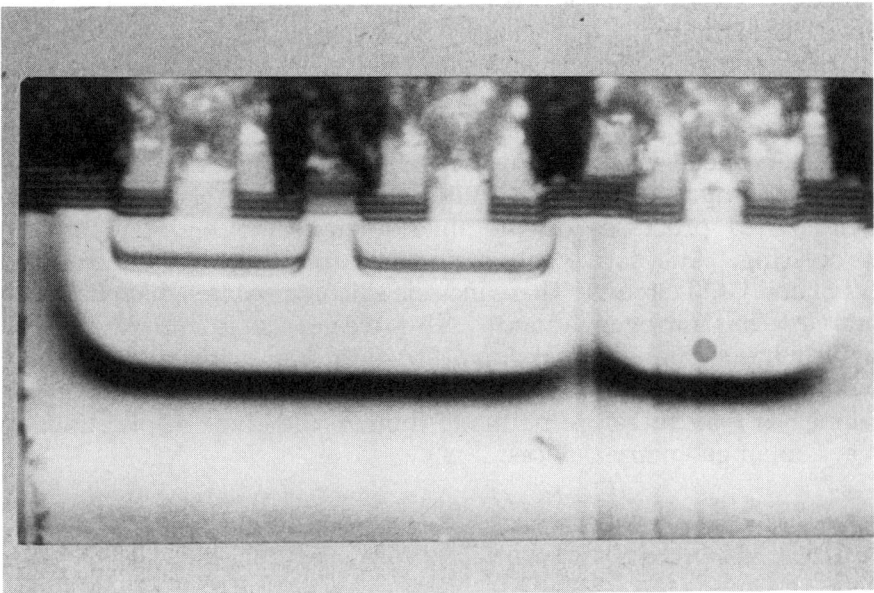

Figure 6.12 Micrograph of a MOSFET (cross section)—source CMD.

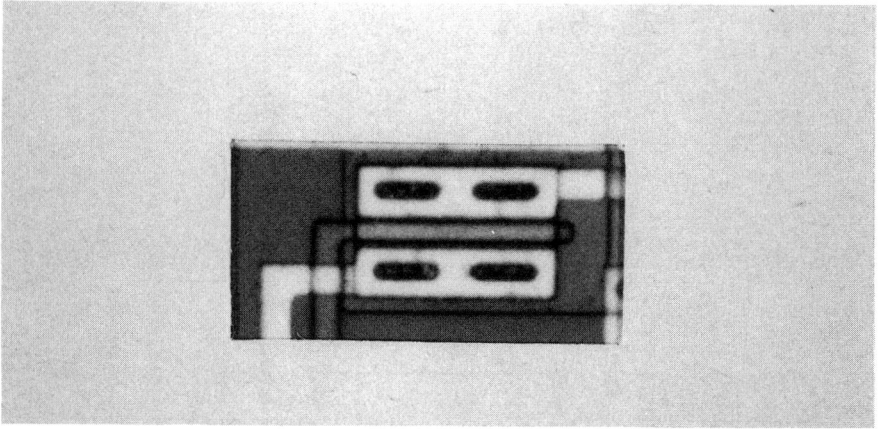

Figure 6.13 Micrograph of a MOSFET (top view)—source CMD.

smaller, that is, have a smaller cross-sectional area, without the total resistance of the interconnect being too high (remember that high resistance combined with capacitance will also slow down electrical signals). Silicide gate MOS is found in high-density, high-speed VLSI and ULSI circuits.

Other gate materials and gate dielectrics have been used for special applications, and new schemes are currently under development. Advances in deposition techniques have allowed refractory metals such as tungsten to be used as gate materials. These conduct almost as well as aluminum but may also be used in self-aligned processes due to their tolerance to high-temperature processing. Alternative gate dielectrics are also in use in advanced VLSI and ULSI circuits. These include silicon dioxide, which has been *nitrided* (has nitrogen added). This creates an *oxynitride* material which has a higher breakdown strength (can withstand a higher voltage) and is more resistant to contamination than silicon dioxide alone. It may therefore be made thinner and finds applications in very small geometry devices.

Other MOS Structures

There are a number of device types based on MOS technology. Some are almost unrecognizable as MOSFETs (as we will see in the next chapter), but others are merely variations on our MOSFET theme. It is this latter category that we will examine in this section.

High-performance MOS (HMOS) uses a technique to make the surface of the semiconductor in the channel more p-type in an NMOS device. This helps to increase the voltage that the device can withstand between the source and drain before the channel begins to conduct. This effect, known as *punch-through,* can be achieved with devices if we increase the source-drain voltage to such a degree that the depletion region of the reverse-biased junction extends across to meet the other junction. The device therefore begins to conduct with little or no voltage applied to the gate. The more p-type the surface of the semiconductor is, the less the depletion region extends across it. The source and drain regions in HMOS may be made closer together without punch-through; hence devices may be made smaller (typically less than 1 micron between source and drain). *Double-diffused MOS (DMOS)* is different from HMOS in that it uses a p-type shield around the n-type source. The shield also acts to prevent punch-through; hence small devices are possible.

One of the most common modifications to the MOSFET is the addition of a *lightly doped drain (LLD)* structure. LDD is used in advanced MOS devices to improve their performance. One of the problems frequently encountered in small-geometry devices is the movement of the source and drain under the gate, particularly during subsequent high-temperature processing. This occurs even in self-aligned devices as the relatively highly doped source-drain regions tend to push dopant sideways, thereby creating gate-source and gate-drain overlap. This also reduces the distance between the source and the drain, thereby encouraging punch-through at lower voltages. In the worst case, it can even short-circuit the source to the drain. The LDD process is somewhat complex, but it results in the reduction of lateral diffusion under the gate by first putting in a very lightly doped source and drain so that the sideways push is lessened. A more heavily doped diffusion is put in afterward, but it is positioned slightly away from the edge of the gate (using reactive ion-etched CVD oxide *sidewall spacers*) so that the electrical resistance of the source-drain regions may be minimized without lateral diffusion under the gate.

The electrical resistance of the source and drain regions is just as important as that of the gate electrode for as device performance. High source and drain resistances slow devices and circuits down and therefore must be minimized. Unfortunately, for the highest-performance devices, heavily doped silicon cannot offer a low enough resistance; therefore the use of low-resistance silicides is necessary. The *self-aligned silicide* or *salicide* process involves putting a transition metal on top of the implanted source-drain regions (on the bare silicon), and simultaneously on the gate poly-Si, and then heating it so that it reacts with the silicon to form a silicide. The metal reacts only with the silicon, not with the silicon dioxide. An oxide sidewall spacer is used at the side of the poly-Si gate to prevent a short-circuit between the gate and the source or drain. The silicide therefore forms only where we wish it—on top of the source, drain, and gate—thereby reducing the resistance of these regions.

Summary

Although JFETs and MESFETs have interesting properties and are relatively simple to make, it is the MOSFET which is predominant in digital ICs. This device operates by controlling the current between two separate p-n junctions by inverting the surface of the semiconductor between them. Depending on the substrate type used, it can be either an n-channel or a p-channel device, the n-channel variant being faster. MOSFETs rely on a high-quality dielectric to isolate the gate (control) electrode from the semiconductor. The MOSFET has extremely good on and off characteristics and therefore is an excellent logic switch for digital applications. A variety of gate electrode materials are used. The most common one in modern devices is polycrystalline silicon, frequently topped with a silicide in advanced circuits. The performance of the MOSFET is enhanced by the utilization of process and structural variations such as the LLD and salicide approaches.

7

Other Device
Technologies

Passive Devices

In order to complete our circuits, the integration of *passive devices* is also necessary. The main passive devices in ICs are *resistors* and *capacitors*. Resistors are used in analog ICs in *voltage dividers* and *current limiters* and are also found in digital circuits as *loads* (see Chapter 8). Capacitors are mainly used in analog circuits to *decouple* dcs between the different stages of a circuit, as parts of *oscillators* and *filters*, and are a vital element in *switched capacitor* systems. Capacitors are also used in digital circuits to reduce *noise*, particularly on the power supply lines, in *charge-pumps* which increase voltages on chip, and to store charge in some types of *memory*.

It is extremely difficult to integrate large-value *inductors* (coils), but it is possible to form small-value components which are ideal for very-high-frequency analog circuits, particularly those which deal with low-power *microwave* signals for communications applications. These *monolithic microwave integrated circuits (MMICs)* require some specialized fabrication techniques to reduce the effects of stray inductances and capacitances, and hence tend to be more expensive to produce than their lower-frequency counterparts.

125

Figure 7.1 shows examples of integrated passive devices. Resistors are generally diffusions in the substrate. Low-value resistors are typically short in length and are formed using heavily doped diffused or implanted regions. These doped areas are of opposite type to the substrate and are usually put in at the same time as the source and drain regions. Higher-value resistors are longer and may be *serpentine* in shape to allow them to fit them into the circuit. Lighter doped regions may also be used for higher-resistance elements, as less dopant naturally means a higher-resistance semiconductor, but the doping cannot be too low; otherwise, the current will flow in the substrate rather than in the resistor, and control of the value will be lost. Therefore, for very high resistance values, we cannot use substrate resistors and must use poly-Si resistance elements instead. Poly-Si which has been lightly doped or undoped has a very high resistance and therefore is ideal for high-value elements.

All we need to create a capacitor are two conductors separated by a dielectric. Integrated capacitors, like resistors, come in two forms. In these elements, the bottom electrode or *plate* is typically a heavily doped area and the top plate is poly-Si. Alternatively, both plates can be made from separate poly-Si layers. In both cases, the dielectric is most likely to be a thermally grown silicon dioxide. To obtain higher-value capacitors, we must either increase the area of the plates or decrease the thickness of the dielectric. Both of these approaches cause problems. The capacitor cannot take up most of the area of the circuit, and a dielectric which is too thin will not withstand operating voltages. Therefore, high-value capacitors are generally not integrated and are added to the circuit externally.

Integrated inductors are typically just surface spirals of metal, as we cannot wind a coil any other way with planar technology. The problem then consists of making the connection to the center of the spiral, particularly for high-frequency circuits. This is solved in MMICs by the use of *air bridges*. A metal connection is made between the center of the spiral and the rest of the circuit. This connection is separated from the underlying coil by a layer of material which can readily be dissolved. Once this supporting layer is removed, the metal connection is self-supporting. This is necessary for microwave circuits, as the use of a permanent dielectric in this position would interfere with the inductor.

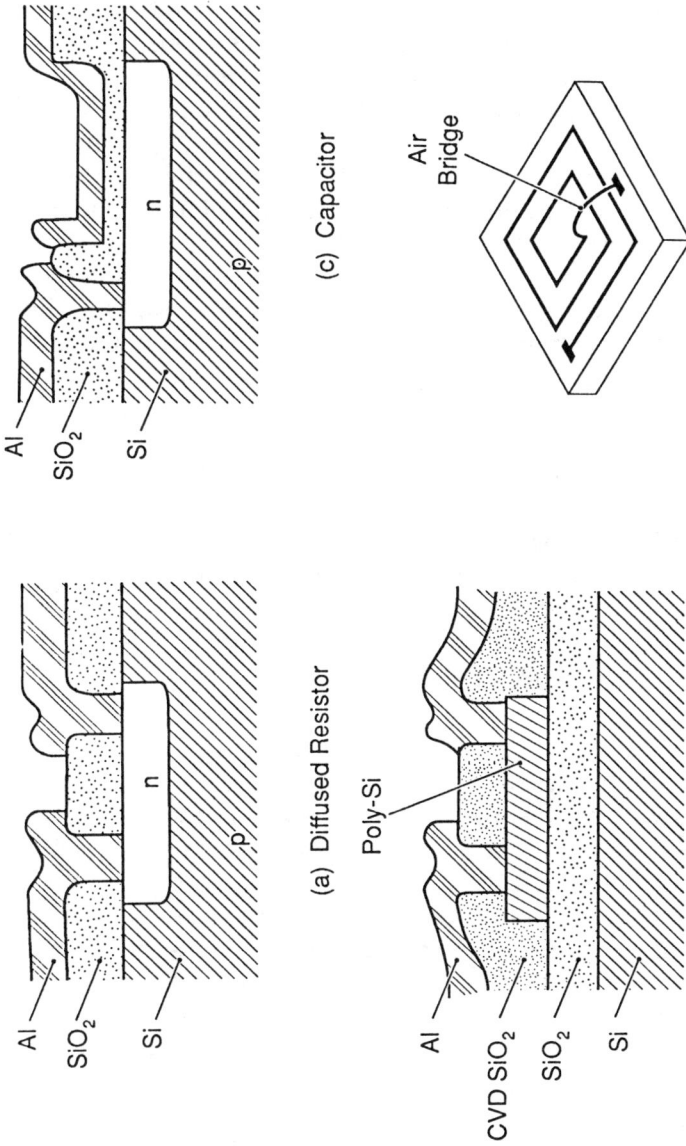

Figure 7.1 Examples of integrated passive components. (a) Diffused resistor, (b) Poly-Si resistor, (c) Capacitor, (d) Inductor.

Diodes

We began our discussion of devices in Chapter 2 by introducing the concept of the p-n junction *diode* and using this to help build our simple models of bipolar and MOS transistors. However, we should be aware that diodes are also important elements within ICs and in electronics in general. As we will see in Chapter 8, small-signal diodes are still extensively used in logic circuits; they are used as *detectors* in analog communications circuits; and high-power diodes are universally found in ac to dc conversion.

Diodes are extremely simple to form using planar technology. An n-type diffusion into a p-type substrate, or vice versa, is a diode. Different types of diodes are employed in ICs. Apart from the normal rectifying junction formed by a heavily doped region in a lightly doped substrate, a *zener diode* can be formed by combining heavily doped regions. In the reverse bias condition, these diodes "switch" at a precise voltage which is designed into the diode by the dopant levels used. These elements can be used as *voltage references* in analog circuits, supplying a constant voltage for applications which require one (voltage regulators, measurement systems, etc.).

High-Power Junction Devices

Some of the differences between small-signal and power transistors were discussed in Chapter 5. However, there is a wide range of power devices which are somewhat different from the BJT and MOS transistors we have examined up to this point. Because of their high current capacity and high power capability, these devices are not often integrated into a circuit and are usually produced as discrete devices. In this section, we will examine the high-power devices which are based on bipolar technology (but are not BJTs as such).

The main high-current devices based on BJT technology are *silicon-controlled rectifiers (SCRs)* or *thyristors* (alternative name), *triode ac switches* (more commonly called TRIACs), and *gate turn-off thyristors (GTOs)*. SCRs are devices which can essentially be in only one of two states—on or off. The structure, shown in Fig. 7.2, is p-n-p-n (like a large-area n-p-n BJT with an additional p diffusion). In the off state, no current flows from the *anode* (the positive electrode) to the *cathode* (the negative electrode), as the central p-n junction is reverse biased. If a voltage is applied to make a current flow into the *gate* electrode,

(A) SCHEMATIC OF STRUCTURE

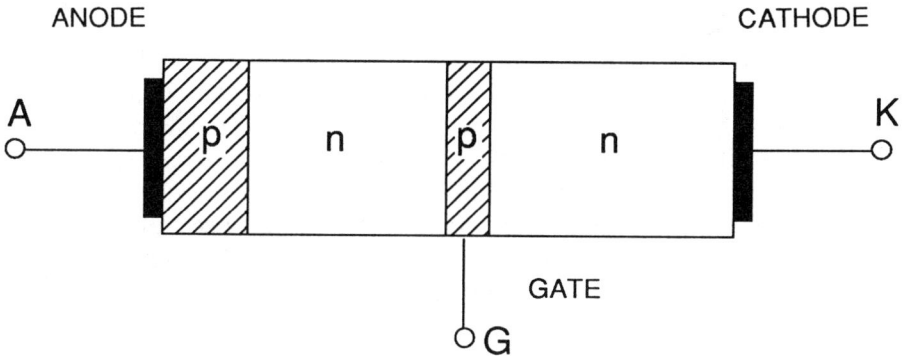

ANODE   CATHODE

A

p   n   p   n

K

GATE

G

(B) CIRCUIT SYMBOL

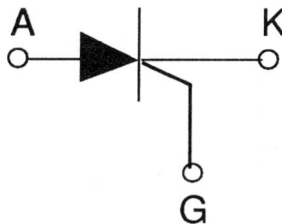

A   K

G

Figure 7.2 Thyristor structure.

the device turns on in the same way as a BJT, but unlike the BJT it cannot be turned off even if the gate current is removed. In fact, the best way to switch the device off is to remove the anode current. This curious latching effect occurs due to the internal biasing conditions that the flow of current sets up. Note that for these devices to operate in this way, the current can be controlled in one direction only (with the anode being positive and the cathode negative, but not vice versa).

TRIACs are back-to-back SCRs and, as such, can control current in both directions. GTOs are thyristors which, like TRIACs, may be turned on with a positive gate voltage but may also be turned off with a negative gate voltage. The negative voltage on the gate removes charge from the second p-type region, and thus conduction through the device may be stopped. The GTO structure is shown in Fig. 7.3.

SCRs and TRIACs are used in alternating current power applications for voltage control. In ac, the current and voltage rise and fall in a periodic fashion. The devices may be triggered in such a way as to let only part of the voltage cycle pass and hence control the effective magnitude of the ac voltage. In ac, the current drops to zero twice in every cycle; hence the devices may be reset (turned off) by this removal of the anode current. GTOs are used in high-power, high-speed switching operations. They can typically withstand higher off voltages than most power devices.

High-Power Field Effect Transistor

There are a number of high-power devices which are based on FFT technology. Amongst these are the *vertical field effect transistor* or *V-shaped groove field effect transistor* (VFET) and the hexagonal field effect transistor (HEXFET).

The VFET is made using a *directional etching* technique to form grooves in the semiconductor. It therefore looks very different from the MOSFETs we have already met (Fig. 7.4). The starting material is generally a lightly doped n-type epi-layer on a heavily doped n-type substrate. The epi-layer is the drain, and the more heavily doped substrate supplies a low-resistance contact to it. The epi-layer then has a p-type region diffused into it, and it is this region which will ultimately form the channel of the device. An n-type layer is then diffused into the p-type region so that we obtain the characteristic back-to-back diode configuration of a transistor. The device at this point appears to be very much like a BJT but we change the situation by performing the groove etch. The groove cuts through the p-type channel layer into the drain. The surface is then oxidized to form the gate dielectric, and the gate metal is deposited and patterned. When a negative voltage is applied to the gate, the p-type region closest to the gate electrode becomes inverted, and a vertical conducting channel is formed between source and drain. Note that these devices are usually used for dc or low-frequency applications where speed of operation is not a consideration and therefore the overlap capacitances (gate-source and gate-drain) are of little concern.

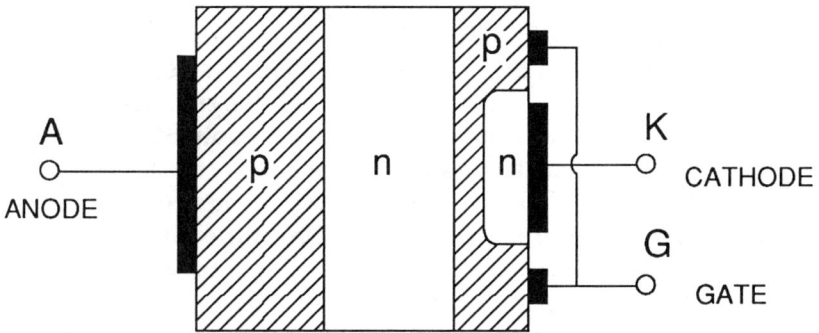

Figure 7.3 GTO structure.

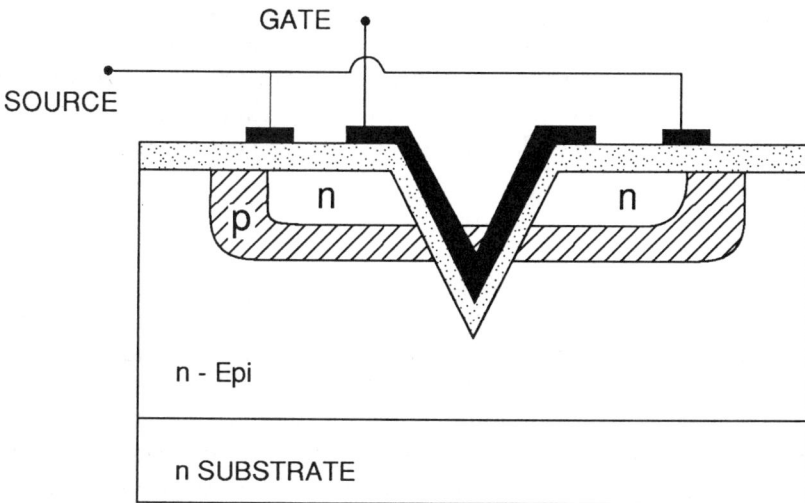

Figure 7.4 VFET structure.

Since the groove can be made very long, a large area for conduction can be created, and therefore the device can have a very low resistance (note that current flows on both sides of the V). This allows it to handle high currents. In addition, the p-type channel region is more heavily doped than the channel of an ordinary

MOSFET, and hence the punch-through effect is minimized. Therefore, the device can also withstand high voltages.

The HEXFET uses an array of hexagonal cells, each acting like an individual FFT (Fig. 7.5). The initial formation steps are generally the same as for the VFET, but the sequence becomes very different after the p-type diffusion is put in. At this point, the surface is oxidized to form the gate dielectric, and a heavily doped poly-Si (or poly-Si plus silicide) gate is deposited and patterned to form the characteristic hexagonal pattern. The patterned poly-Si is then used as a surface mask to self-align the n-type source diffusion or implant. The net effect is that the poly-Si covers the p-type diffusion where it meets the surface (both are separated by the gate oxide, of course), but the source regions are open, so that a blanket metal connection may be made to them after the poly-Si is insulated by a suitable CVD oxide. The channels which form when a negative voltage is applied to the gate occur at the surface just under the gate, so that the current first flows horizontally from the source and across the channel, and then turns to go vertically into the drain. This happens in each of the cells, so that the cumulative current carrying area can be very large.

Both devices are somewhat similar to DMOS, but due to their geometries, they may be packed closely together to achieve a large current-carrying area. In fact, a suitably packaged HEXFET may carry as much as 100 amps.

Thin Film Transistors

Thin film transistors (TFTs) are typically MOSFET-type devices based on a semiconductor which has been deposited on an insulator (Fig. 7.6). Insulators such as silicon dioxide are generally *amorphous*, which means that they are disordered or noncrystalline. Since it is extremely difficult to form a single-crystal material on an amorphous substrate, the semiconductor itself is generally amorphous. This leads to poor conduction (through low carrier mobility) and leaky p-n junctions, so the device characteristics are nonideal. Therefore, TFTs tend to be used in applications requiring simple switches (on insulators), such as the control elements in transparent *electro-optical displays*.

The device shown in Fig. 7.6 has a typical TFT design. The amorphous silicon has been etched to form islands on the insulating substrate. The junctions are diffused or implanted at either end of these islands, and the metal contacts to these junctions are put on

GATE
POLY Si

SOURCE METAL

SOURCES ARE
INTERLINKED
BY METAL

DRAIN CURRENT

DRAIN

n

p

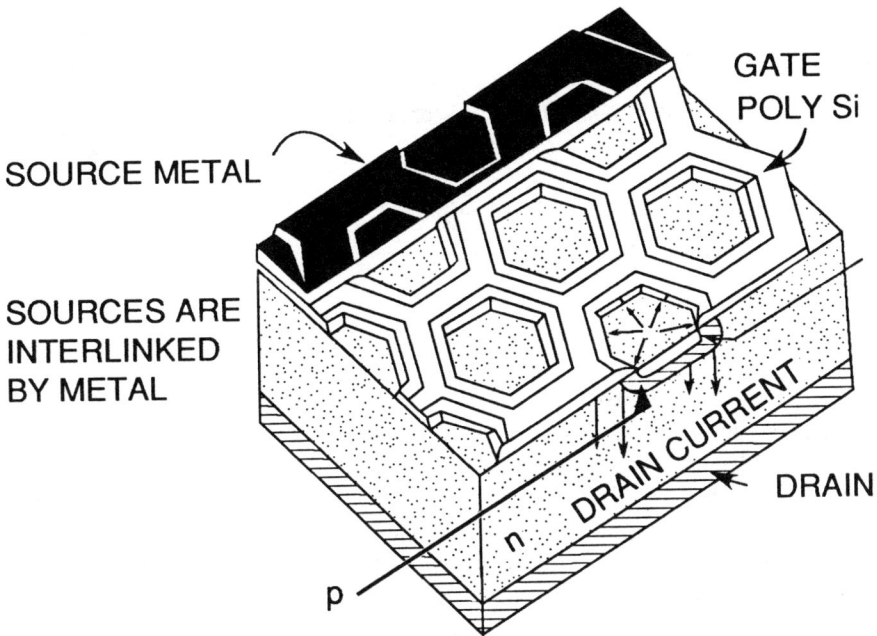

Figure 7.5 HEXFET structure.

prior to the deposition of a gate dielectric. Note that a deposited gate dielectric has poor characteristics, but this is of little consequence in the TFT, as the substrate material is so poor anyway. The gate metal is then deposited and patterned, and the device is complete.

Microwave Devices

Microwave devices are small-signal devices which must operate in the frequency range 10^8—10^{12} Hz (remember that 10^{12} Hz means 1 million cycles per second!). This has traditionally been the realm of highly specialized vacuum tubes such as the *magnetron* and the *traveling wave tube (TWT)*. Semiconductor devices such as the *tunnel diode*, the *metal-insulator-semiconductor (MIS) diode*, the *metal-insulator-metal (MIM) tunnel transistor*, and the *impatt diode*, now have a firm hold in this frequency range. These devices are not far removed from what we have studied so far, but they are able to handle such high frequencies

SEMICONDUCTOR   GATE   INSULATOR

INSULATING SUBSTRATE

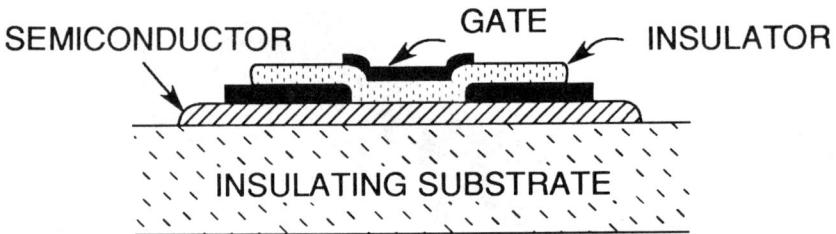

Figure 7.6 Thin film transistor.

by the use of extremely small depletion widths or insulator thicknesses. However, it is the heterojunction bipolar transistor (HBT) which promises to be the microwave switching element of the future, replacing much of the current technology, particularly in monolithic microwave integrated circuits (MMICs). MMICs are revolutionizing data processing and communications technology but, as mentioned previously, are expensive to produce.

Nonvolatile Memory Structures

As we will see in the following chapter, we can create digital memory circuits for computer applications using BJTs or MOSFETs. However, the main problem with memory circuits formed using this approach is that the circuits "lose their memory" if the power is removed. Fortunately, a number of *nonvolatile memory (NVM)* device technologies are available. Nonvolatile elements can retain information in electrical form even when the electrical power supply is removed. Note that we can also store information in a nonvolatile fashion on *magnetic media* such as magnetic *tape* or magnetic *hard or floppy discs*, but these systems are external to the main processing circuits of the computer. What we will examine in this section are nonvolatile elements which can be incorporated into ICs. The two main technologies are really just modifications of MOS devices. These are *floating gate* and *metal-nitride-oxide-semiconductor (NMOS)* transistors.

A schematic of a floating gate NVM transistor is shown in Fig. 7.7. As may be seen, this device is very similar to a MOSFET except that it has two gate electrodes, the poly-Si middle or *floating gate* being completely isolated electrically by layers of oxide below and above. When a large positive voltage is applied to the top or *control gate,* electrons are forced to pass through the thin oxide between the substrate and the floating gate by a process known as *tunneling,* thereby accumulating in this gate. Since the gate is electrically isolated, the electrons can be retained in it for a long time without significant leakage (ten years is typical). The process can be reversed by placing a significantly large negative voltage on the control gate to push electrons back through the tunnel *oxide* and into the substrate. When the floating gate is *charged* with electrons, it creates an electric field which puts the surface of the substrate into inversion and thereby switches the device on. With no charge on the gate, there is no electric field and the device is in the off state. We can therefore store digital information in electrical form.

MNOS devices store the electronic charge not in a floating gate but at the *nitride-oxide interface* (Fig. 7.8). The interface between a thermally grown oxide and a CVD nitride has many faults or *traps* where the atoms of the insulators do not bond together completely. Electrons may be stored in these traps. The interface traps may be filled or emptied electrically as before, or they may also be emptied by shining ultraviolet light on the device, which also causes the electrons to pass through the oxide.

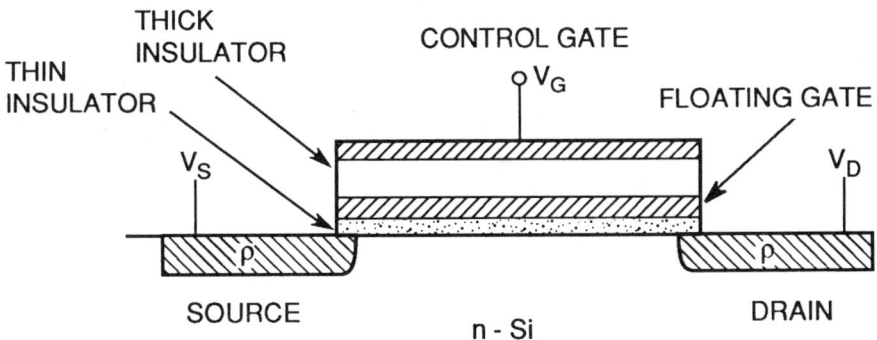

Figure 7.7 Floating gate NVM.

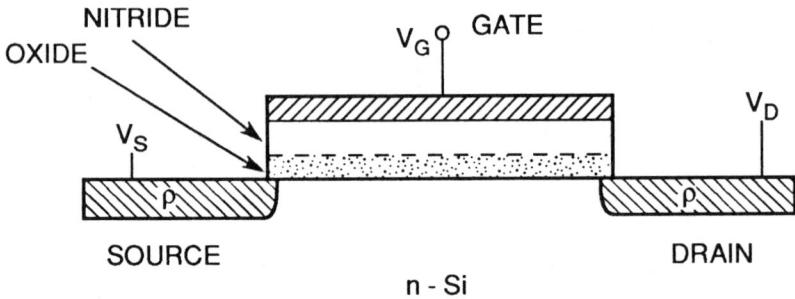

Figure 7.8 MNOS structure.

The programming of these devices requires relatively high voltages to get the electrons to tunnel through the oxide. The higher voltages are typically generated on chip by a circuit consisting of capacitors and diodes called a *charge-pump*. One of the big drawbacks of the charge storage technologies for military or space applications is that they are not very *radiation hard*. If they are subjected to ionizing radiation, they may become discharged.

Other NVM technologies use magnetic materials to store information on chip rather than in charge storage devices. Currently, the most widely used of these alternative technologies is *bubble memories*. These store information magnetically in a *garnet* substrate and may be programmed and read electrically. Unfortunately, their use requires a special package with a coil incorporated; this, along with the expense of the substrates, makes bubble memories costly. *Ferroelectric* materials such as *barium titanate* appear to be an exciting alternative for future NVM schemes. These materials can store information, as their *dielectric constant* can be electrically altered. If they are used in a capacitor structure, this change may be electrically detected. They may thus be programmed and read electrically. The effect is reversible, so they may also be reprogrammed. Both of these information storage schemes are radiation hard and are therefore of great interest to the military.

Integrated Injection Logic

Integrated injection logic (I^2L) is another bipolar technology variant. I^2L is effectively the bipolar equivalent of CMOS, as it involves (lateral) p-n-p transistors and (vertical) n-p-n devices fabricated together to save area and make the full device compact.

Figure 7.9 shows a cross section of a typical I^2L device. A lightly doped n-type epi-layer has been grown on a p-type substrate containing a heavily doped n-type buried layer. As in all modern BJT devices formed using planar technology, the buried layer reduces the resistance to current flow in the subsurface epi-layer circuit, as the resistance of the epi-layer alone is high. However, in this case, the epi-layer and the buried layer are the common emitters of the vertical n-p-n devices (these are effectively upside down compared to the BJTs we have seen before). The multiple collectors are the last to be diffused or implanted in I^2L.

This rather odd-looking layout allows the compact combination of p-n-p and n-p-n devices, but the devices are also configured to be ideal for digital logic circuits and memory designs. Its main advantages are that it is compatible with BJT processing and it has a higher packing density than the normal BJT, as the guard rings around each transistor may be omitted. In fact, the connections of the devices through the epi-layer are deliberately used (rather than prevented) to enhance the compactness of the full device.

Charge-coupled Devices and Surface Acoustic Waves

There are many applications, particularly in *digital signal processing (DSP)*, where the controlled *delay* of a digital or analog signal is required. There are two basic device types which can supply a time delay. These are discussed here.

The first and more common type is the *charge-coupled device (CCD)*. It is basically an array of closely spaced *MOS diodes* (Fig. 7.10), or it may be viewed as a very long MOSFET with a gate which has been split into multiple parts. In this device, digital information is represented by a *packet of charge*, which is merely a bunch of electrons. This packet is put into the device at the input electrode by applying a (negative) voltage pulse there. With a negative voltage on the first gate, the packet is launched into the channel, but the channel is not complete across the length of the device. However, under the application of a sequence of negative *clock pulses* (regular timed voltage pulses), the charge packet can be transferred along the device by means of a moving depletion region under the clocked gates into which the charge will flow. This is often referred to as the *bucket*

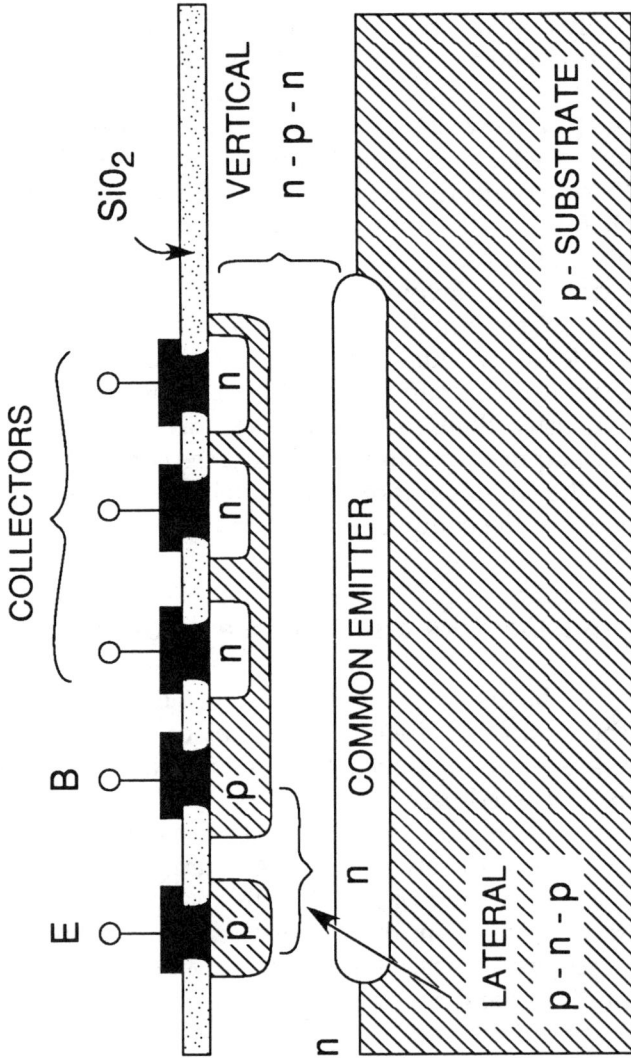

Figure 7.9 I²L schematic.

brigade effect. Since the speed of transfer of information is determined by the speed of the clock pulses, a variable delay is possible.

This effect is useful in many information processing applications involving digital signals, but CCDs are also sensitive to light (as are all semiconducting devices), and so they are finding applications in optical systems and photodetectors as well. If light hits the channel as the packet is being moved across, it can alter the amount of charge which reaches the other end of the device; hence a CCD array can be used for imaging. In fact, the newer lightweight, hand-held video cameras use CCD transducers to convert the light image into an electrical signal which can be recorded on magnetic tape. The resolution of these devices depends on how many gates we have per unit of area.

Another type of *delay line* is the *surface acoustic wave (SAW)* device. This uses a *piezoelectric* material on the surface of a substrate. This type of material will compress or expand when a voltage is applied. When an electrical signal is applied to the input, an *acoustic wave* (a small shock wave) travels across the surface of the device. The speed of this wave depends on the acoustic velocity in the material. A delay therefore exists between the send and receive points, and the length of the delay can be determined by the length of the path it has to travel (once the acoustic velocity is known). Since the amplitude of the wave depends on the magnitude of the applied signal, different sizes of waves may be launched; therefore this device may be used for analog signal processing applications. However, digital signal processing is rapidly taking the place of many analog systems, as SAWs are considerably more expensive than CCDs.

Solar Cells

Solar cells are very simple semiconductor devices which use light energy to produce an electrical current. The device generally consists of an n-type semiconductor substrate which may be single-crystal silicon, amorphous (disordered) silicon, or various compound semiconductors, with a thin p-type diffused layer on top to create a large area p-n junction.

When light strikes the cell, electrons and holes (which we normally call *electron-hole pairs*) are created at the junction. These are swept away by the internal voltage, electrons into the p-type region and holes into the n-type region. Since we have generation and

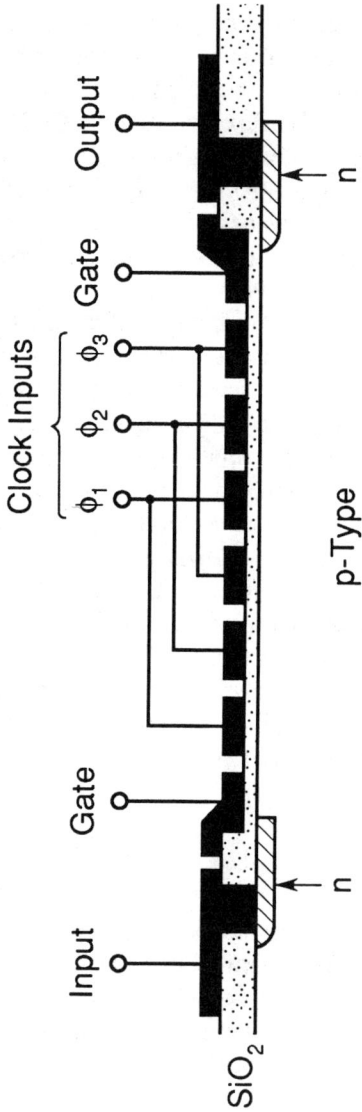

Figure 7.10 CCD structure.

movement of carriers, a current flows, and as long as we have light energy falling on the cell, we can extract power from it.

Photovoltaic power systems have found extensive applications in recent times, but if the scheme sounds like "energy for nothing," prepare for a disappointment. The most efficient cells are those based on single-crystal compound semiconductors with multiple epitaxial layers. This allows much of the light which falls on them to be converted to electrical power, but they are extremely expensive and therefore tend to be used only for space vehicles. Single-crystal silicon is fairly efficient and its cost is lower, but the energy needed to refine and grow the silicon is usually more than can be obtained from the cells during their normal working life. The least expensive alternatives, in terms of the energy used to make them and their dollar cost, are those based on amorphous or polycrystalline silicon. These are also the least efficient, as the defects in the semiconductor tend to swallow up many of the generated carriers, but they can be made into very large area *solar panels* and hence show great promise for the future for domestic use (particularly if the price of oil rises beyond acceptable limits). Most commercially used solar cells are currently found in applications which require photovoltaic power for convenience (rather than being more energy efficient than an alternative source). Such applications include solar-powered calculators, walkway lighting systems, and fans.

Light-emitting Diodes, Lasers and Detectors

The *bandgap* of many compound semiconductors, such as indium phosphide, is such that when carriers (electrons and holes) recombine at a p-n junction, visible light is emitted. Similarly, light of a particular wavelength entering the junction can create electron-hole pairs, similar to what happens in a solar cell. These structures may then be used as *light-emitting diodes (LEDs)* or *photodiode detectors* which give off and respond to particular wavelengths of light, respectively. We can also use the light emission property of semiconductors in special *heterostructures* to create semiconductor *lasers*. Normally, lasers are large units containing rods or tubes, but semiconductor lasers are monolithic and may be made extremely small. They can even be integrated into compound semiconductor based circuits.

LEDs are used as *indicators* instead of light bulbs in panel displays, whereas lasers and detectors are used in *fiberoptic* communications. This latter application is particularly interesting, as electrical signals can be converted to pulses of light and these pulses sent along thin glass fibers instead of sending electrical signals down long wires. The benefit of this conversion is that we no longer have to worry about capacitance and resistance slowing our signals down; hence both the speed and the amount of information we can send are greatly increased. The current state of the art in optoelectronics involves information processing by transistors as usual, but with lasers and photodiode detectors at the outputs and inputs, respectively. However, the future of information processing on chip will doubtless lie with optical processing, with optical elements doing the job of transistors and data being carried by integrated optical waveguides.

High Electron Mobility Transistors and Superlattices

The *high electron mobility transistor (HEMT)* is a high-speed device based on GaAs technology. Other names are the *selectively doped heterostructure transistor (SDHT)* and the *modulation doped field effect transistor (MODFET)*. A schematic representation of this device is given in Fig. 7.11. It is similar in some respects to a MESFET, except that the substrate contains buried source and drain regions, and the overlying epi-layer has an undoped region between the source and drain. The name heterostructure comes from the combination of the different layers in this fashion. The very high speed of the device is achieved by the high mobility of electrons in undoped GaAs. The electrons are provided from the AlGaAs donor layer, and once they enter the undoped region, they may move very rapidly across the device from source to drain.

A further development in semiconductor technology is the *superlattice*. A vertical superlattice uses alternating thin layers of semiconductors of different bandgap such as GaAs and AlGaAs or silicon and germanium. An alternative form is the lateral superlattice, which is made using very fine metal lines on the surface of a semiconductor to create an array of closely spaced depletion regions. These so-called *quantum devices* confine electrons in a particular way and are examples of what is now called *bandgap engineering*. They provide an artificial lattice effect (hence the name superlattice) and the

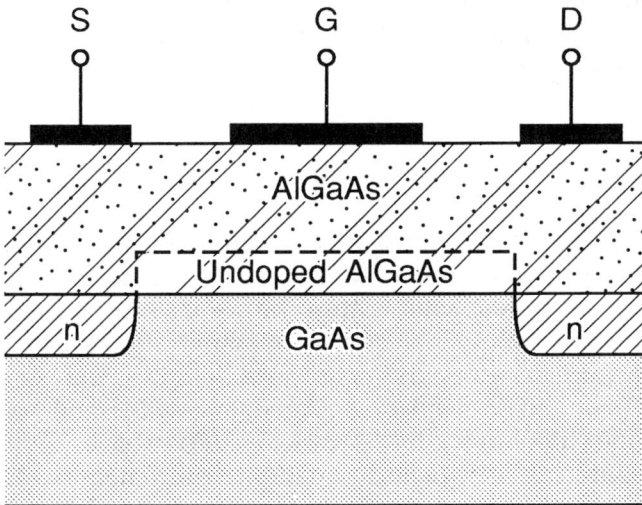

Figure 7.11 HEMT structure.

physical characteristics of the device may be altered by the period of the superlattice. For example, the effective bandgap of the superlattice may be tuned to a specific application in optical signal detection by fabricating a lattice with an appropriate period. In some respects, this is like creating an artificial semiconductor with designed-in properties.

Combination Technologies

Many circuits now employ combinations of device technologies. This enables parts of circuits which perform specific tasks to use the technology most suited to each task. Examples of these combinations are:

CMOS-NMOS. High-density, high-speed NMOS is used for memory areas, and low-power CMOS is used for peripheral (input/output) and control circuits.

BiMOS or BiCMOS. Data processing is performed by NMOS or CMOS circuits, but the output drivers are high-power BJT devices.

GaAs on Si. This emerging technology uses high-speed or optical elements in GaAs with silicon data processing circuits. The GaAs layer can be formed on top of the silicon substrate by molecular beam epitaxy techniques.

Summary

Although BJT and MOS transistors dominate in ICs, there are many other component types which exist either as discrete devices external to the IC or as co-residents within the monolithic circuit. Passive devices are particularly necessary, as many ICs, especially analog systems, require resistors and capacitors. High-power devices such as HEXFETs and SCRs can be integrated, just as BJT and MOS can be united in one chip, but the larger varieties of power components exist as discretes due to their heat dissipation requirements. Many types of display use LEDs or thin film transistors formed in a deposited semiconductor, and CCD chips are almost universally found in modern lightweight camcorders for imaging. Semiconductor lasers and detectors are being used in communications and may one day may be applied to practical ultrafast optical computing schemes. Nonvolatile devices now have a vital role to play in memory systems which retain information when the power to the system is lost, and are more frequently found in complex microcontrollers.

8

_____

Circuits and Systems—
A Cross Section

Digital Circuit Concepts

In this chapter we will take a brief look at the types of circuits which use the basic MOSFETs and BJTs we examined in Chapters 5 and 6. This is not a totally comprehensive study but rather an overview to illustrate how these devices are used. We will begin by looking at the basic circuit concepts for digital applications.

The basic logic circuit configuration is the driver-load arrangement shown in Fig. 8.1. The configuration is shown for the three basic logic gates: NOT (inverter), NAND (not AND), and NOR (not OR). The logic symbols for the gates are also included in the figure. Using integrated components, the NAND and NOR logic functions are easier to implement than the corresponding AND OR functions. In Fig. 8.1, the logic functions have been realized using switches to represent the transistor drivers and resistors as the loads. The inputs in this case are merely mechanical linkages to the switches, so that we may open or close them, open representing logic 0 on the input, closed representing logic 1. One of the functions of the load resistor is to limit the flow of current from the upper *power supply rail* to the lower one (usually at zero volts) when the switch is in the closed position.

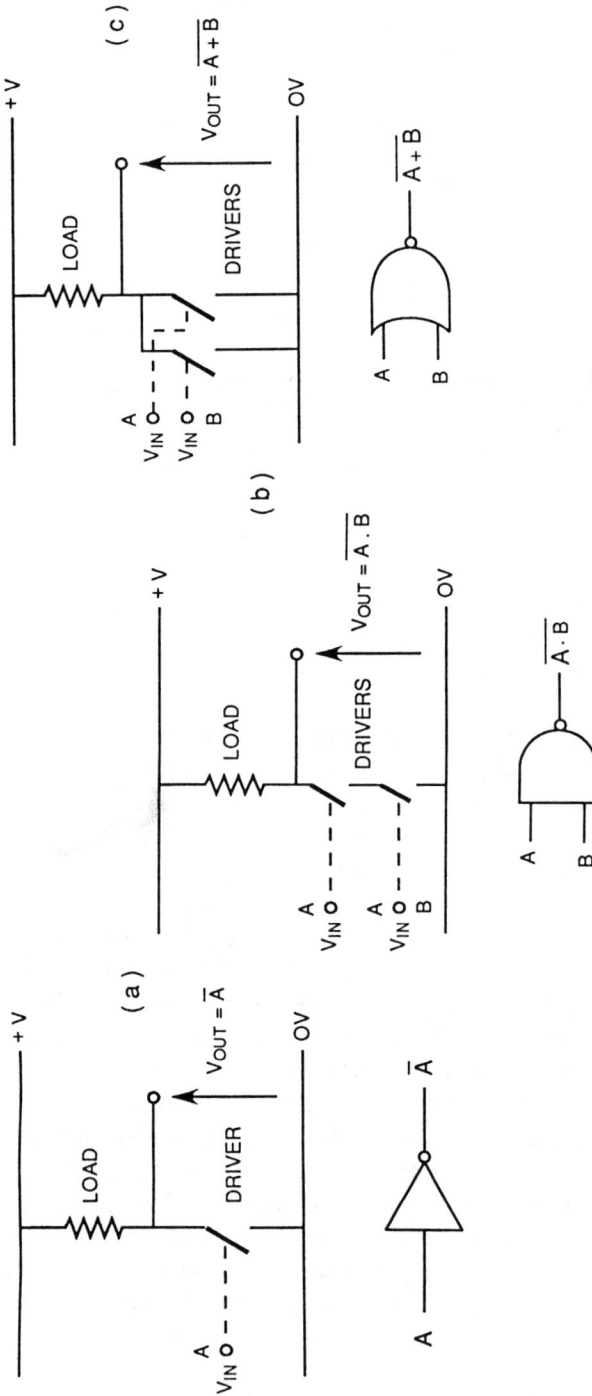

Figure 8.1 Realization of logic functions using switches. (a) NOT gate or inverter, (b) NAND gate, (c) NOR gate.

The output is the voltage measured between the lower power supply rail and the point below the load.

This driver-load configuration is merely a *voltage divider* which splits the full rail-to-rail voltage arithmetically between the load and the driver; the voltage drop across a component is proportional to the resistance of that component. If the resistance of the driver is much higher than that of the load, most of the voltage will appear across the driver and hence will be measured at the output. If the driver has a very small resistance compared to the load, a smaller portion of the power supply voltage will appear at the output.

In the NOT case, when the driver is off (switch open—logic 0 on input), the resistance of the driver is much higher than that of the load (for an ideal switch, the off resistance is infinite). Therefore, the output shows a high voltage which corresponds to logic 1. When the driver is on (switch closed—logic 1 on input), the opposite is true and the output is low, which corresponds to logic 0. The NAND and NOR implementations operate by the same principle, but multiple inputs are involved. In the NAND case, both switches have to be closed before the input is low. If one or both are open, the input remains high. Similarly, in the NOR case, both switches have to be open for the output to be high. If one or both are closed, the output will be low. As we can see, these gates perform the logic functions described in Chapter 1.

In the NAND and NOR gates, we may increase the number of inputs by adding more switches in series or parallel for the two functions, respectively. In real-logic ICs, the switches are, of course, transistors. However, it should be kept in mind that transistors do not behave exactly like ideal switches; their off resistance is not infinite, and their on resistance is not close to zero.

Digital MOS Circuits

MOS technology is frequently used in digital circuits, as MOSFETs are extremely good switches. They typically have better switch-like characteristics than comparably sized BJTs, exhibiting a lower on resistance and a higher off resistance. As discussed in Chapter 5, they may also be packed more closely together; this is an advantage in complex logic circuits (modern microprocessors have over 1 million transistors!). Unfortunately, due to the capacitance created by the gate oxide, digital MOS circuits tend to be somewhat slower than digital bipolar versions, but the latter circuits inevitably consume more power.

In a MOS driver-load configuration, we usually employ a *depletion-mode MOSFET* as a load, rather than using a large-length resistor. The driver is an *enhancement-mode MOSFET*. A typical depletion load inverter is shown in Fig. 8.2. The gate of the depletion-mode device is connected to its source to provide appropriate *biasing* conditions. This saves a great deal of area, as one transistor is considerably smaller than a long serpentine resistor and provides ideal load characteristics, giving a sharper division between logic 0 and logic 1 than a resistor (or, for that matter, an enhancement-mode MOSFET load, which is also used in some cases when depletion-mode loads are not available on chip). Note that the power supply voltage for most commercially available logic circuits is five volts dc (this is why we require power supplies to convert the 115 volts ac which comes to our offices and homes).

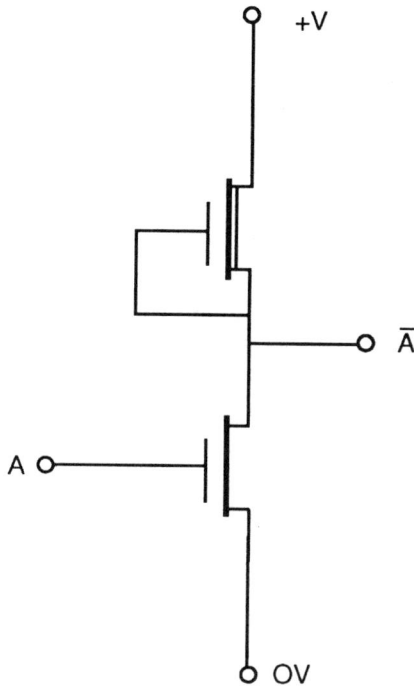

Figure 8.2 Depletion load inverter.

When we apply (close to) zero volts to the input (corresponding to logic 0), the driver is turned off. The resistance of the enhancement-mode driver is higher than that of the depletion-mode load; hence most of the voltage across the two transistors appears across the driver. Because of the way the gate of the depletion-mode device is connected, it is biased in an on condition and therefore is forced into a relatively low-resistance state. This is further insurance that most of the supply rail voltage is dropped across the driver and hence appears at the output as logic 1. Note that we say "close to" and "most of" when we talk about the voltages corresponding to the logic levels. Due to the threshold voltage effect, there is a little latitude in input voltages. As far as output voltages are concerned, due to the finite off and nonzero on resistances of transistors, we can always expect to be a little off zero volts for logic 0 and not quite at five volts for logic 1.

When we apply (close to) the supply rail voltage to the input (corresponding to logic 1), the driver is turned on. The resistance of the enhancement-mode driver is lower than that of the depletion-mode load; hence most of the voltage across the two transistors appears across the load. Once again, because of the way the gate of the depletion-mode device is connected, it is biased to be in a relatively high-resistance state (being a depletion-load device, it is still a little on and therefore can supply some current to the output). This is further insurance that most of the supply rail voltage is dropped across the load, leaving little to be dropped across the driver; hence a logic 0 appears at the output. We thus have our inverter characteristic implemented in a compact fashion using two transistors. Note that we may use PMOS or NMOS in this configuration, although NMOS is more common in microprocessors and other VLSI logic circuits due to its inherent speed advantage.

The problem with MOS logic—and with bipolar logic, for that matter, as we will see—is that when the driver is on, some current flows from rail to rail even with a depletion-mode load in its relatively high-resistance state. Therefore the overall current flow and power consumption of MOS circuits are high (they can be even higher for a bipolar logic gate). In CMOS logic, there is no load as such. Instead, we have two enhancement-mode transistors, one n-channel and the other p-channel. The gates are connected together to form the input of the inverter, as shown in Fig. 8.3. If we apply a (relatively) positive voltage to the input (corresponding to logic 1), the n-channel device turns on but the p-channel device turns off. This makes the

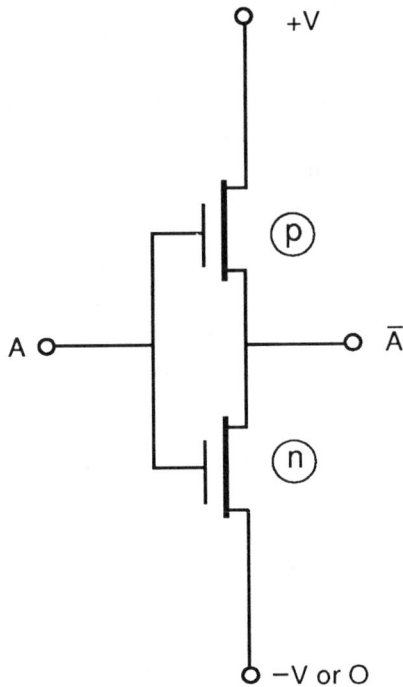

Figure 8.3 CMOS inverter.

output go near to the lower supply rail voltage (corresponding to logic 0). If we apply a (relatively) negative voltage to the input (logic 0), the p-channel device turns on but the n-channel device turns off. This makes the output go close to the upper supply rail voltage (logic 1). (Note that we can get the same effect if the upper supply rail and logic 1 are a positive voltage and the lower supply rail and logic 0 are at zero volts, as zero volts is relatively negative compared to a positive voltage.) Therefore we can see that this acts as an inverter, but, more importantly, only one device is on at any one time.

This is what makes CMOS a *low power consumption* technology: current never drains through the device when the driver is on, as it does in NMOS or PMOS (except in the brief moment when the devices actually switch). We therefore find CMOS circuits wherever there is a need to conserve power—for instance, if the circuit has to run off a small battery, as in watches, calculators, or even laptop computers. Also, since the *quiescent* current is low in CMOS, the amount of heat generated in each part of the circuit is low; therefore we can pack more gates together without having to worry about them overheating severely. CMOS can also be useful for analog applications, for instance, signal processing. However, it is not as suitable for applications like amplification, as the *transfer characteristics* are not usually as good as those of BJT technology and *distortion* of the signal can occur as a result. Whereas NMOS dominated in the 1970s for integrated logic applications, CMOS became the industry standard in the 1980s and BiCMOS, the combination of CMOS and bipolar, is the key IC technology of the 1990s.

We have seen how MOSFETs may be combined to form simple logic gates. To end this section, we will take a brief look at digital MOS circuits which contain more than two transistors. The first, shown in Fig. 8.4, is a three-input NOR gate implemented in CMOS. There are six transistors in total in this gate, three n-channel drivers and three separate p-channel loads. This configuration provides the best implementation of the function, with well-defined output logic levels. The other circuit we will look at is the CMOS *flip-flop*, illustrated in Fig 8.5. The flip-flop or *bistable* is often used to *latch* (hold) a *bit* of data. The circuit is configured in such a way as to stay in a selected logic state. For instance, putting a logic 1 on the set "S" input, even for a moment, makes the output "Q" go to logic 1. When this happens, the other output "Q bar" goes to logic 0 and, in turn, holds Q at logic 1 even when the stimulus on S is removed. Similarly, if the reset "R" input has a logic 1 applied while the flip-flop is in the "set" state, the output Q goes to logic 0 and Q bar to logic 1, once again remaining in this state even when the high at the input is removed. We have created a simple memory circuit which is often used in groups to hold multiple bits of data at the input of a digital circuit until the circuit accepts it for processing.

In digital design terms, the logic gates are examples of *combinatorial* logic circuits in which the outputs are determined solely by the instantaneous logic values at the inputs. The flip-flop is a simple example of a *sequential* circuit, in which the output depends not only on the inputs but also on the previous values of the inputs. Both types of circuit are found in microprocessors and other complex logic-based systems.

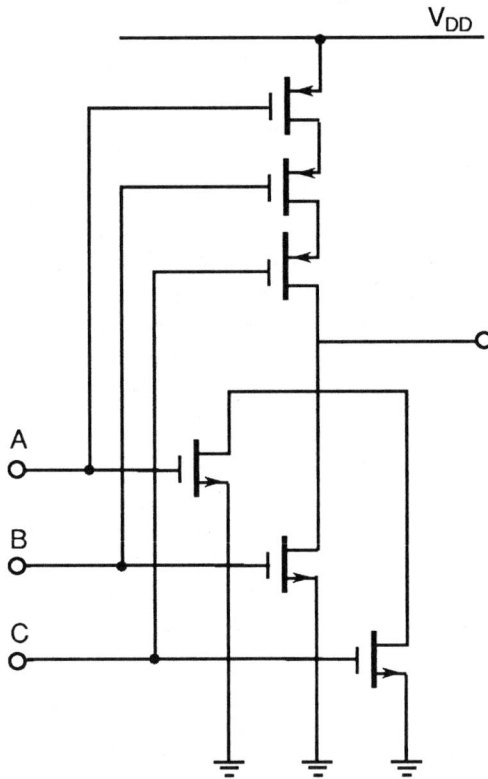

Figure 8.4 Three-input CMOS NOR gate.

Digital Bipolar Circuits

Logic functions realized in bipolar circuitry have been commercially available almost since the first days of the IC. They have appeared in various forms, all generally based on the driver-load arrangement discussed previously. However, the BJT is not as good as the MOSFET for digital applications (except that it can be faster), and hence the circuits tend to be a little more complex. The main configurations are as follows:

DTL (diode-transistor logic). This technology uses a combination of diodes and transistors to implement logic gates. A typical DTL NAND gate is shown in Fig. 8.6. If either of the two inputs is close to zero volts (logic 0), point C is held low and the transistor is off, leading to a logic 1 on the output. However, if both inputs are at a positive voltage (logic 1), point C can go high and the transistor turns on, leading to a logic 0 at the output. This is a very simple implementation of the NAND function, but DTL is not particularly popular in modern bipolar logic circuits because simpler implementations are now available.

TTL (transistor-transistor logic). This is quite similar to DTL, but the input diodes are replaced by a *multiple-emitter* BJT as shown in the simple circuit of Fig. 8.7. The input transistor is biased appropriately by the base resistor so that it effectively replaces three of the diodes in the DTL version. TTL is easier to fabricate and more compact than DTL. Most of the off-the-shelf, simple combinatorial circuits use TTL. It also has the advantage of being electrically robust, as we can add a relatively heavy current output stage implemented using large BJTs.

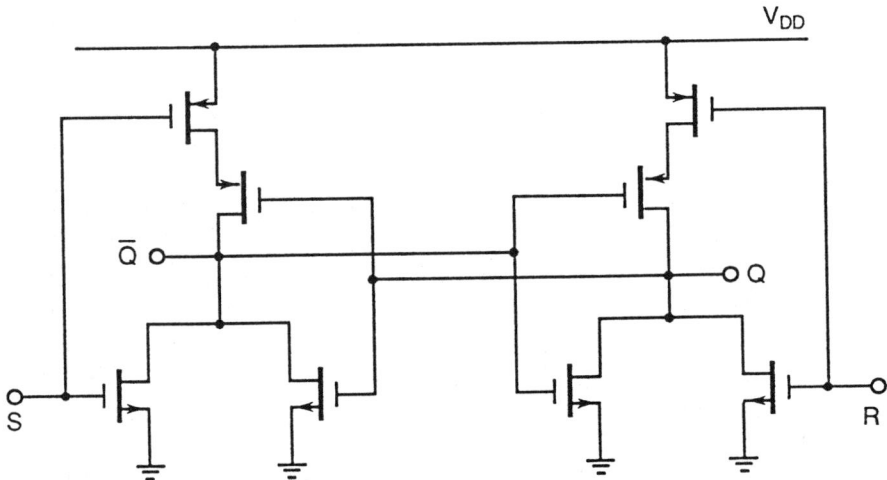

Figure 8.5 CMOS SR flip-flop.

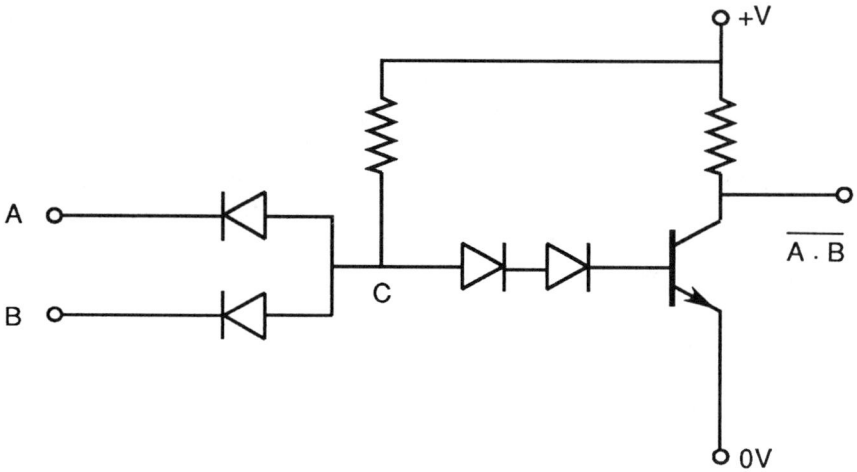

Figure 8.6 DTL NAND gate.

Schottky TTL. This is similar to TTL, except that the input is a *Schottky transistor* in which the input p-n junction is replaced by a *rectifying metal-semiconductor contact*. This is much faster than TTL, as Schottky junctions do not have *minority carrier recombination*. TTL is widely used in fast SSI logic circuits.

ECL (emitter coupled logic). This technology is very fast, as it uses a configuration which does not allow the transistors to *saturate*. This basically means that the transistors are operated in the region where they do not switch hard on; they are effectively kept on the edge of switching. This allows them to switch rapidly when required. A typical ECL inverter is shown in Fig. 8.8. The biasing resistors ensure that the transistors do not saturate. ECL is widely used in large, fast mainframe computers. Unfortunately, due to the fact that ECL draws considerable current, systems using this fast logic often require forced cooling to deal with the large heat generation.

We have discussed the speed advantage of bipolar logic, but the technology has one further advantage over MOS logic. Since we can add relatively large BJTs at the output, bipolar logic typically has a better *drive capability* or *fanout*. This means that the output can be

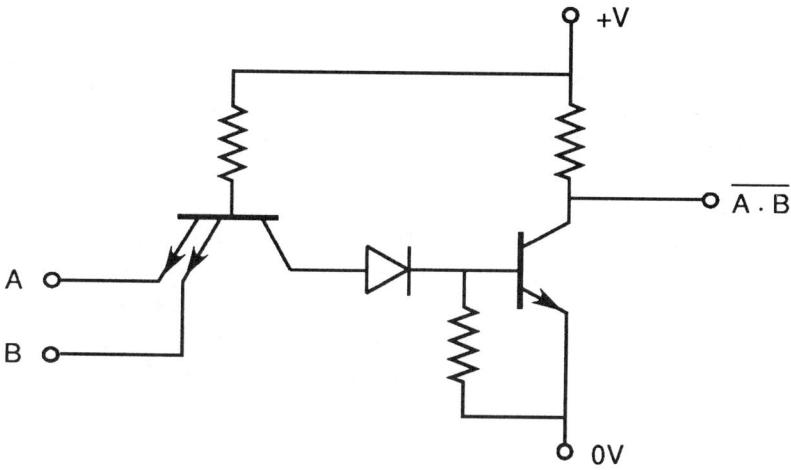

Figure 8.7 TTL circuit.

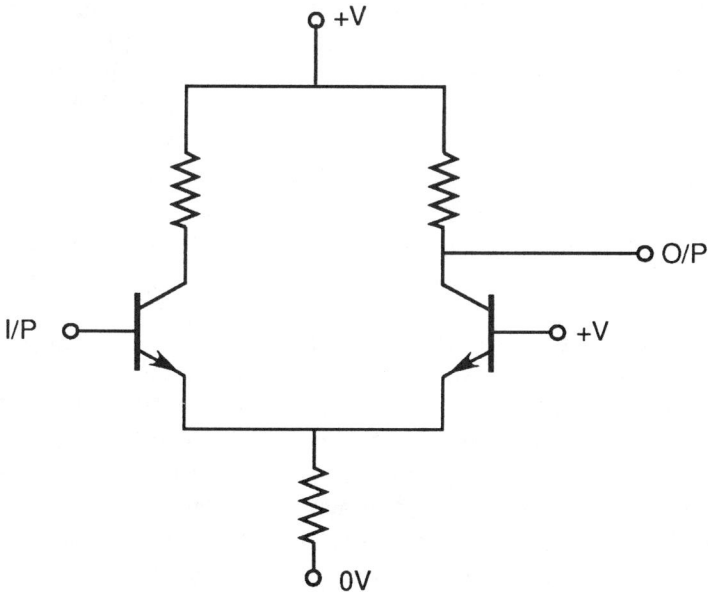

Figure 8.8 ECL inverter.

connected to many devices, including the input stages of several other logic gates in the system, without overloading the circuit. MOS logic is not as good in this respect, but the combination of MOSFETs and BJTs in BiCMOS circuits has helped to alleviate this problem with MOS logic.

Unfortunately, as the diagram of the two-input TTL NOR gate of Fig. 8.9 shows, the actual realization of bipolar logic tends to be more complex than that of MOS logic. The bias resistors and the added output drivers tend to make the physical area of the gate very large,especially when we add structures like guard rings around the transistors; therefore the packing density is very poor. This, along with the current consumption and heat dissipation problems, is why there are no bipolar logic equivalents to the microprocessor on the market. Such circuits tend to be only found in special packages in the heart of large computers with dedicated cooling systems.

Analog Bipolar Circuits

At the risk of overgeneralizing, let us note that MOS transistors are more suitable to logic applications in modern ICs, whereas bipolar devices are probably better in many respects for analog applications. This is due to the way the outputs vary with respect to the inputs for the two types of transistors. Currently, we are more likely to find BJTs in a typical analog chip, but this trend is slowly reversing as we develop new ways to use MOSFETs in an optimal fashion in analog ICs, which are also called *linear* circuits.

A number of basic analog circuits are implemented using BJTs. They include the following:

Amplifiers. These circuits which use transistors to amplify a small analog electrical signal into a large signal in stages. A typical amplification stage is shown in Fig. 8.10. The output current from the previous stage flows into the input transistor and causes the current through its output circuit to change. This, in turn, drives the larger output transistors (which can carry a larger current), which provide the main amplification of the current. Note that both n-p-n and p-n-p devices are used. The values of the resistors are chosen to give appropriate biasing for the transistors so that they provide the most *linear transfer characteristic*, that is, the output is a linear function of the input.

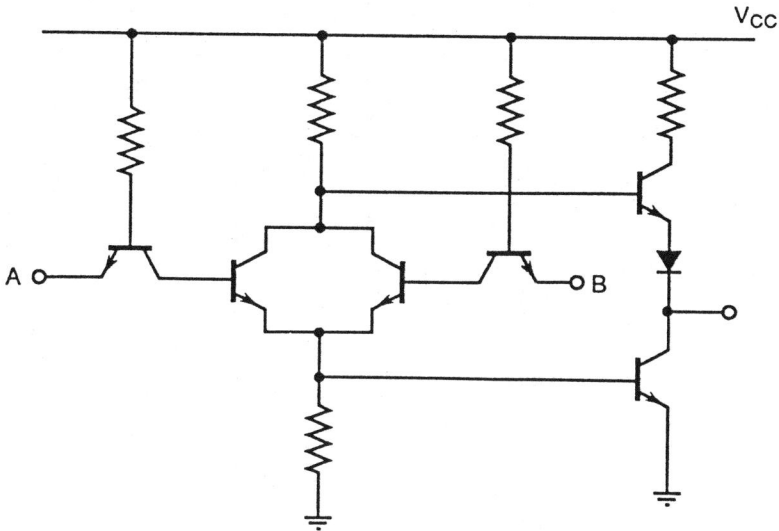

Figure 8.9 Two-input TTL NOR gate.

Each stage may use progressively larger devices to achieve higher levels of amplification. Figure 8.11 gives the circuit diagram of a typical *operational amplifier* (op-amp), showing an input stage (a *differential* stage in this case, which amplifies voltage differences), a gain stage, and an output stage. The upper n-p-n transistor of the output stage pulls the voltage up, and the lower p-n-p pulls the voltage down. Op-amps such as this are used in many analog signal applications, as well as to implement analog functions such as *differentiation* and *integration*. Audio amplifiers are similar to this design, except that they may have more intermediate stages and no differential input stage.

Oscillators. These are used to generate frequencies (audio frequencies, radio frequencies, etc.). The configuration generally is a resonant circuit consisting of resistors and capacitors or inductors and capacitors linked to an amplifier. The frequency of oscillation is determined by the values of the components in the resonant circuit. A typical oscillator, utilizing an op-amp, is shown in Fig. 8.12.

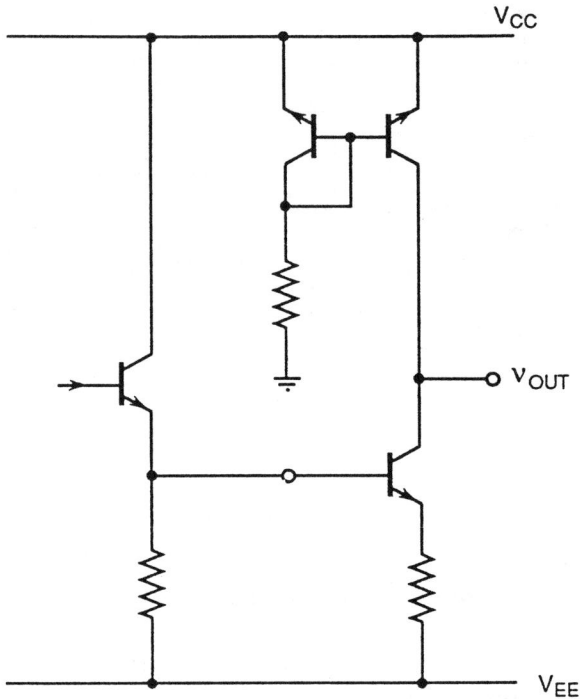

Figure 8.10 BJT gain stage.

Filters. These circuits use resonant circuits at the input to an amplifier to block or admit selected frequencies. They are used as tone controls in audio amplifiers and tuning circuits in radio. A typical filter, utilizing an op-amp, is shown in Fig. 8.13.

Analog MOS Circuits

Amplifiers using MOS technology often use it in the input stages, where a *high input impedance* is important. The MOSFETs replace the input transistors, but the intermediate gain and output stages use bipolar devices. However, in recent times, more MOSFET-based amplifier designs have appeared, such as the one shown in Fig. 8.14. This is very similar in many respects to the bipolar op-amp, but there

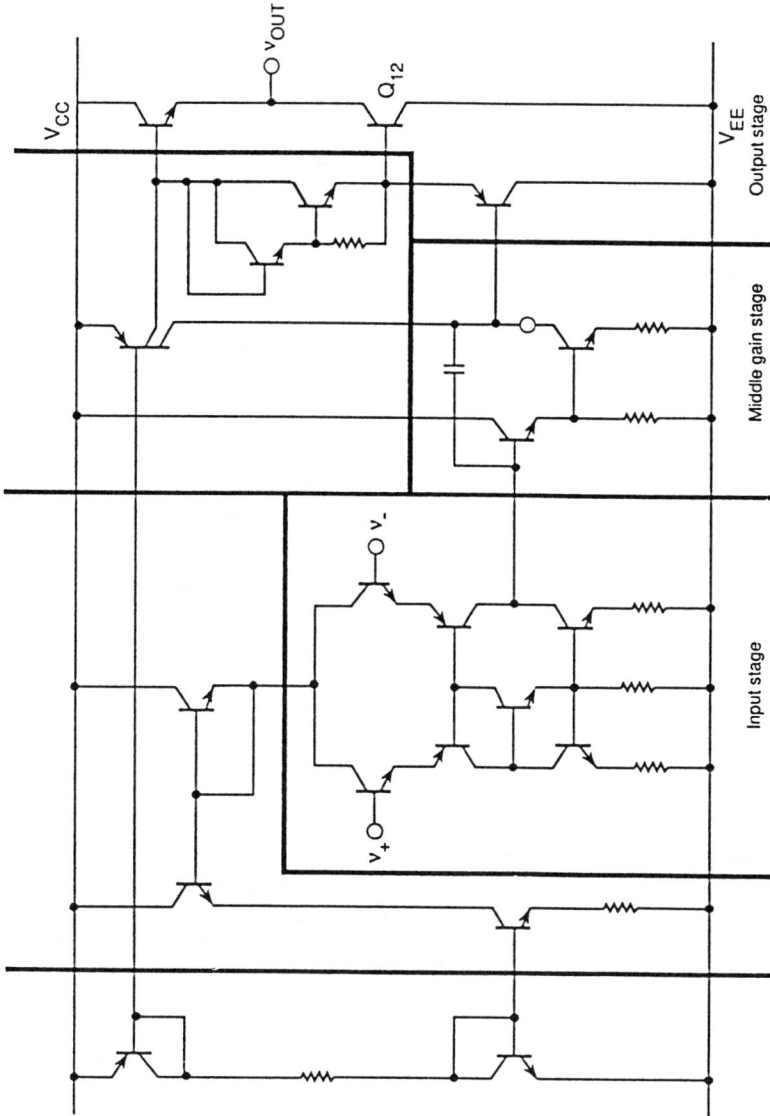

Figure 8.11 Assembled stages of an op-amp adapted from Horenstein (see Further Reading).

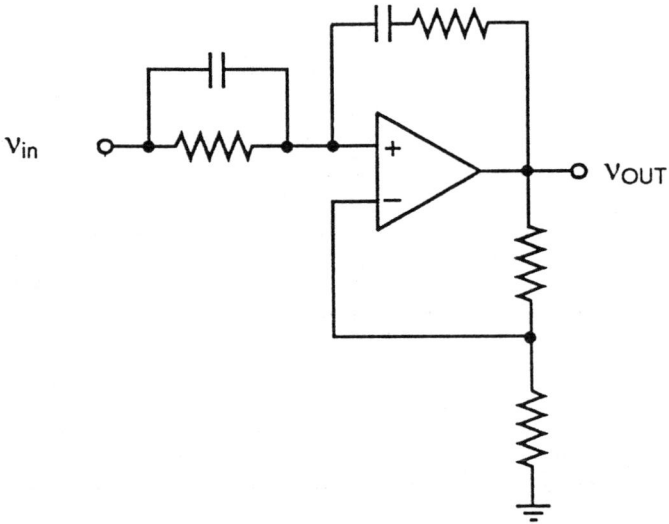

Figure 8.12 Oscillator circuit with an op-amp.

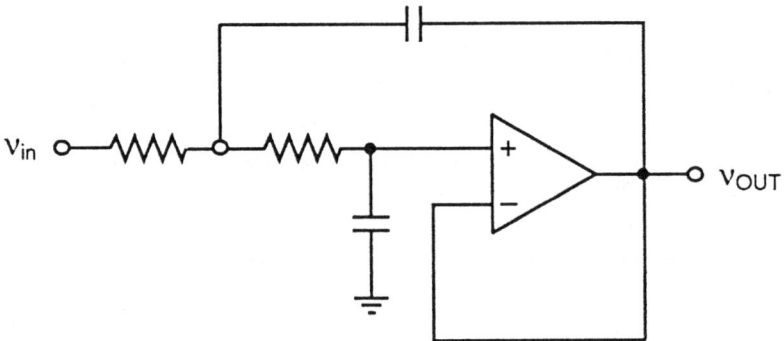

Figure 8.13 Filter circuit with an op-amp.

are no bias resistors as such. Biasing is provided by enhancement-mode MOSFETs which have a constant gate voltage applied. As in digital MOS, the use of a transistor as a high-value resistor saves a great deal of area and provides more precise characteristics.

Linear CMOS has also become common, frequently using two layers of poly-Si to create capacitors for filtering applications. We can also synthesize the operation of filters by using *switched capacitor* circuits in which MOS devices switch voltages onto capacitors (the theory is a bit difficult, but the net result is a circuit with precise filter qualities).

Microprocessors and Memories

Microprocessors and other types of single-chip computers invariably utilize the high packing density and low power of NMOS and CMOS. Microprocessors contain the necessary *sequencers, registers,* and *arithmetic and logic units (ALUs)* to perform data manipulation and control functions. To create a computer system, *program* and *data memory* must be added. Some single-chip microcomputers have limited memory, input/output devices, and *keyboard* and *display* drivers on the same chip as the ALUs. MOS microprocessors are not as fast as ECL computer chips, but single-chip ECL computers are not common due to power dissipation and packing density restrictions. Figure 8.15 shows the main subsections of a typical microprocessor.

Electronic memories take many forms but essentially fall into two major categories: *volatile* and *nonvolatile*. In volatile memory, the information stored is lost when the power is removed, whereas nonvolatile memory can retain information with no power applied. A typical volatile memory is *random access memory (RAM)*. This is the usual type of memory used for short-term data storage in computers.

RAM has two forms: *dynamic RAM (DRAM)* and *static RAM (SRAM)*. DRAM can store information only for a fraction of a second, even with power on, and therefore it has to be *refreshed* or rewritten regularly. However, DRAM elements may be made very small, and therefore many millions of bits of information may be stored on a single IC. The most basic DRAM element is merely a MOSFET with a capacitor connected to its gate. The capacitor holds charge (until it leaks away) and hence can be used to retain information for a short time. In modern high-density DRAMs, the capacitor is the gate of the MOSFET and no other capacitor is required. However, since this capacitance is small, *sense amplifiers* must be used on chip to detect the state of the storage element. SRAM has a lower packing density and

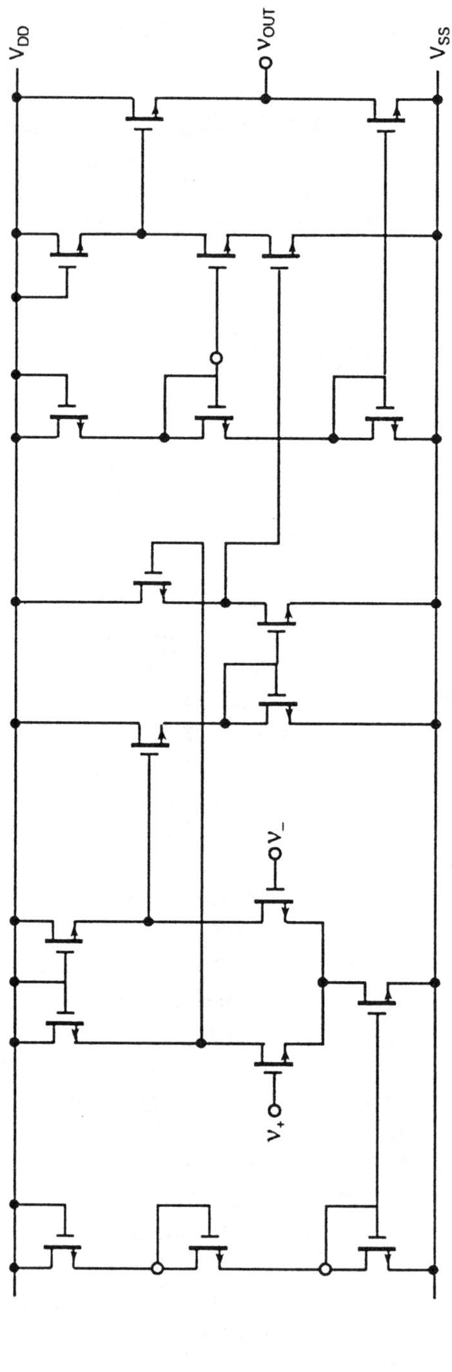

Figure 8.14 NMOS op-amp circuit adapted from Horenstein (see Further Reading).

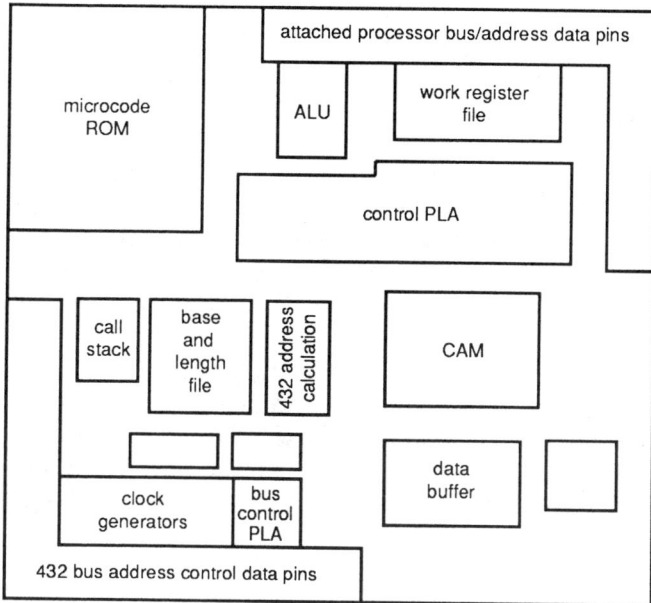

Figure 8.15 "Floor plan" of a typical microprocessor.

requires many more transistors, but it does not require refreshing. It frequently uses the cross-coupled configuration we saw in the flip-flop to create a circuit which is stable in one of two states (logic 0 stored or logic 1 stored). The load or *pull-up* resistors in SRAM are frequently very-high-resistance poly-Si resistors.

A typical electronic nonvolatile memory is *read-only memory (ROM)*. This is the type of memory used to store programs and data for systems which may lose power. ROMs are programmed during manufacture by custom designing the final metallization layer. In this way, we create the necessary connections between arrays of diodes, or more typically, transistors (usually MOSFETs) to permanently build in the desired *code* (program or data). An alternative is the user *programmable ROM (PROM)*, which may be programmed electrically by blowing *fusible links* in the circuit. These weak links are put in the final metal layer, so that the user of the PROM can selectively remove them by passing a high current through those which have to be broken. The connections to the devices in the array are thereby defined, and the code is permanently programmed in. Both these types cannot be reprogrammed once programmed.

ROMs which may be reprogrammed are *erasable PROM (EPROM)* and *electrically erasable PROM* (EEPROM or E2PROM). The former may be programmed electrically, as it utilizes the nonvolatile devices we examined in Chapter 7. However, EPROMs must be irradiated with ultraviolet light to erase all the data before reprogramming. EPROM packages are equipped with windows to allow the light to reach the circuit. EEPROM, which also uses nonvolatile devices in a particular circuit configuration, can be completely altered (reprogrammed) electrically and does not require ultraviolet light for erasure.

As mentioned previously, other memories used in electronic systems are magnetic (magnetic tape and disk, bubble memories using ferromagnetic materials) and *optical* (laser disk). These types (with the exception of bubble memories, which can be like ICs) have a much higher information capacity than semiconductor memories but ICs are catching up.

Summary

A huge number of circuits and systems exist in integrated form. The main categories of circuit are analog and digital, both of which may be implemented using BJT or MOS devices. MOS is particularly suitable for digital applications, but BJTs are used if high speed is necessary. Most NMOS digital circuits use a depletion-mode load, but CMOS, which has no load as such, has recently become the dominant device technology in logic ICs. The key digital systems are microprocessors coupled with memories. The main memory types are ROM, RAM, and EEPROM. Analog circuits, such as op-amps, have been mainly bipolar, but more MOS-based designs have been appearing steadily since the late 1970s.

9

Design and Test

Design Concepts

Design of an IC within any technology adheres to a set of *rules* which provide information on device characteristics and determines how to place circuit elements relative to one another. For instance, the design rules give basic information on how small we can make devices, how close together we can put them, and how we can interconnect them. These rules are necessary, as the process used to create the circuits creates limitations on, for example, the smallness of the device elements may be and how close together they may be placed. If these rules are not followed, the finished circuit may not operate adequately due to unwanted device-device interactions or device failures.

The choice of technology, that is, which process should be used, depends very much on what the resulting circuit has to do. For instance, a battery-operated, low-power circuit would use CMOS, a high-speed computer circuit would use bipolar ECL, and so on. Each technology has its own design rules and usually its own specialized designers. There are different design rules for technologies with different *linewidths*, the smallest defined dimension within the circuit. For example, the design rules for a 5 micron linewidth NMOS process are very different from those for a 1 micron NMOS process. The smaller geometry process is chosen if speed and packing density are concerns, as smaller linewidths mean smaller devices. The larger 5 micron process would most likely be used if cost is important, as equipment for larger geometry processes is less expensive and the cost of the production facilities can also be considerably less.

Design with a few tens or even a few hundred components can be done by hand on a drawing board. The design can then be transferred by hand or by machine to a large mask, which would be subsequently *photo-reduced*. For a circuit containing a few million devices, hand design is out of the question and complex computer systems are used to aid the designers.

Computer-aided Design

The modern IC design system is based on a computer which runs a number of software packages to aid the design process. There are two basic categories of design software: *synthesis* and *analysis*. Synthesis programs help the designers create the actual shapes of the devices and interconnects, and lay out circuit elements to form the complete integrated system. Analysis programs are used to determine how the finished device will perform before it is actually fabricated. Digital circuit design lends itself well to the use of design aids, as there are not too many rules and most of them are clear-cut. Analog design is much more difficult; and hence some of the design software has become available only since the advent of expert systems. The main types of programs within the two categories are discussed below.

Synthesis Programs

Graphic aids. These allow *symbolic representation* of the circuit under design, either at *circuit* or at *logic representation* level. They are used to create an electrical diagram of the design using circuit or logic symbols, or to create a physical *layout* which resembles the actual IC, complete with the final shapes of the transistors and other components. In the latter case, the circuit is built up device by device and layer by layer in the computer. The designer builds up each device by placing, for instance, the gates and source-drain regions in the appropriate positions and interconnects them with contacts and

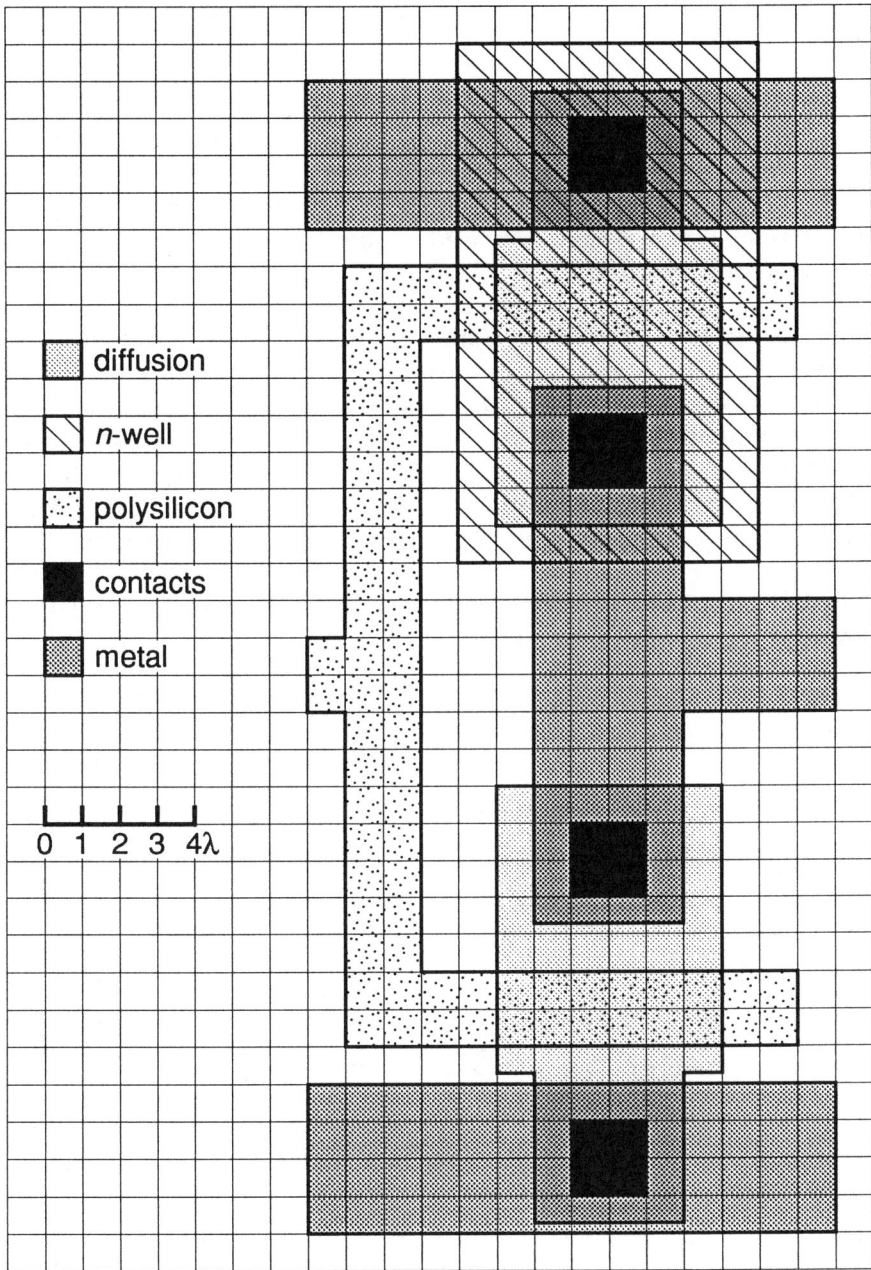

Figure 9.1 CMOS inverter layout.

metal layer wiring patterns. Figure 9.1 is an example of a layout for a CMOS inverter.

Autorouting. In a complex circuit, the interconnecting of devices could take many months of designer time. Programs exist which interconnect functional blocks of devices. The input to these programs is often a software electrical representation of the circuit. The interconnection design is carried out according to the design rules for the technology used.

Symbolic artwork. To further optimize designer time, design may be achieved by using a program to translate a symbolic electrical layout, such as the one shown in Fig. 9.2, to the actual circuit layout of Fig. 9.3. The symbolic layout can be a traditional circuit diagram, showing all the components in symbolic form, or a stick diagram which basically shows how the various devices have to be connected. The physical layout is "fleshed out" from these inputs automatically.

Compilation. This technique enables the designer to input to the computer the details of how the circuit must operate. The computer

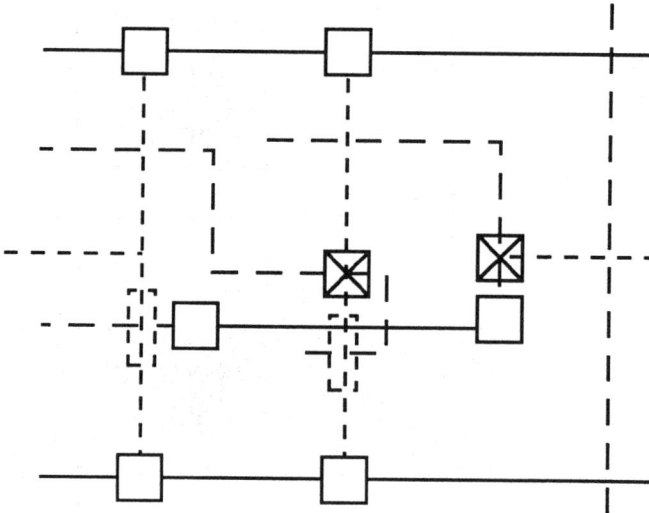

Figure 9.2 Stick diagram of a layout.

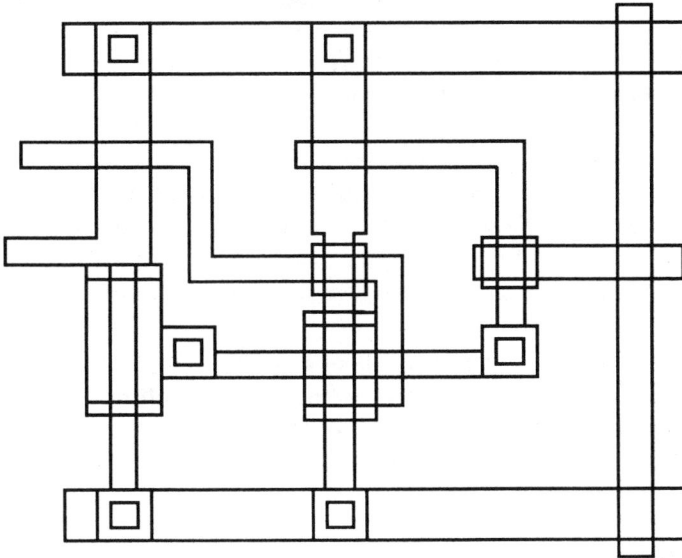

Figure 9.3 Actual layout of a circuit.

translates these concepts into a finished IC layout, without the need for the intermediate design of an electrical layout. This may make it seem as though human IC designers are totally redundant, but these programs are currently limited to fairly simple designs and tend to be available only for digital circuits.

Analysis Programs

Design rule checking. This program checks the finished layout for infringements of design rules. It checks all the dimensions within the finished layout against the design rules for the technology in question and draws any errors to the attention of the designer.

Circuit extraction. This program checks to see if a finished design layout will actually perform the desired logic function. It works at logic gate level to determine if the functionality is correct, but it will not reveal if the circuit will have problems due to, for example, parasitic elements.

Simulation. Simulation programs are the most advanced types of analysis software. They can take the designed layout and the known electrical characteristics of the devices made by a particular process and use them to "test" the circuit in full without the need to actually fabricate the circuit. They require information taken from real devices in order to build up a software model of all the components of a circuit. This is done by running a *test chip* or a *process verification module (PVM)* which contains individual devices through the production process. The result is then analyzed to extract the relevant device parameters. This saves a great deal of time and money if there are mistakes in the basic electrical design or component layout.

The design of modern complex VLSI and ULSI ICs, such as microprocessors and memories, would be unthinkable without computer-aided design. In effect, computers are involved in the design process so that we may design more complex computers!

Full Custom Design

Full custom design is the most complex, time-consuming, and costly design method. It is used for high-volume product lines which require complete optimization of circuit parameters and maximum packing density. Currently, this is best achieved by seasoned designers who can use their experience to optimize the layout so that no space is wasted and to ensure that each section of the circuit functions as close to perfect as possible. This normally takes tens of man-years for the larger designs. Typical full custom circuits are microprocessors and other large digital and analog integrated systems.

Full custom design involves the design of the circuit from the bottom up. No part of the circuit would normally be a stock-designed element. However, many designs, although in the realm of full custom, have previously designed sections which can be lifted from other full custom circuits.

Semicustom and Application-Specific Design

Full custom design takes a great deal of time and effort and is therefore extremely expensive. Only a small proportion of designs need to be produced this way; the other, less exacting designs, which make up the bulk of all new circuits, do not have to have the same degree of optimization. The alternative is *semicustom design*.

This form of design uses *predesigned elements* which are interconnected to form the finished circuit. It is a faster and cheaper design method. Semicustom design is usually employed for low- to medium-volume product lines where optimal packing density or circuit performance are not essential. Semicustom design falls into two broad categories: *gate array* and *standard cell*.

Gate arrays, also called *programmable logic arrays (PLAs)*, *uncommitted logic arrays (ULAs)*, and a variety of proprietary names such as *Master Slice* and *Gate Forest*, use a nearly completed circuit as their base. This base consists of a matrix of identical units, such as groups of devices which may be connected to form logic gates. The stage missing from the complete circuit is the final level of interconnection. This is the only layer which has to be designed, since the other (lower) layers have already been designed and fabricated. The metallization is the only layer which is specific to the application. This layout connects selected elements to form the final circuit. The technique has a quick *turnaround* time but is wasteful of *real estate* (the available area for circuit on the chip), as it is unlikely that all devices can be used in a circuit.

The standard cell approach uses a library of basic functional elements. The elements may be simple logic gates or complex arithmetic units. The best libraries can have hundreds of cells. The end product circuit is designed by linking selected cells together at the circuit layout level. The technique is more flexible than the gate array approach but takes longer, as the entire circuit has to be fabricated from the ground up.

Current Trends in Design

Current trends in microelectronics indicate a definite move toward *application-specific* or *application-optimized designs*. *Application-specific integrated circuits (ASICs)* are one of the fastest-growing market sectors in microelectronics. At present, *commodity* circuit designs dominate. Commodity chips, which range from simple logic gates to microprocessors, are produced in large volumes for multiple applications (the applications can come after the circuits are designed to some extent). In contrast, ASICs are produced for particular applications (and even for single customers). They may be produced using full custom or semicustom design methods.

The computer-aided design packages are shifting the design effort away from the design specialist to the engineer who needs the circuit. It is now possible for an engineer unqualified in IC design to design a circuit based only on the knowledge of how the circuit must perform. However, full custom designers are always required to create the cells for cell libraries and to design complex circuits which must be completely optimized.

In addition, most of the above techniques apply only to digital design. Analog design is more complex due to a wider array of circuit parameters. Therefore, analog designers will always be required as long as there is a need for totally analog circuits.

Testing and Testability

Once a device or circuit is produced, there is no guarantee that it will work. As we will see in Chapter 11, a component may fail to operate for many reasons. We are therefore obliged to test components to assess their worthiness. It is relatively simple to test a simple component, such as a single resistor, capacitor, or transistor. A set of voltages or currents are applied to the device and measured. Alternatively, signals of different frequencies are applied, and the magnitude of the output is measured. If the results fall within a predetermined range, one which fits the desired design parameters, then the device passes the test.

The more complex a circuit is, the more difficult it is to test. If a particular logic circuit can have only ten predetermined outputs for certain inputs, then it can be tested fairly rapidly, especially using an *automatic test system*. However, a considerably more complex circuit may have thousands or even millions of *output states*. Testing for all of these outputs would be too time-consuming; therefore different test methods would have to be employed.

For complex circuits which may have many output states, a *testing algorithm* must be used which exercises only certain key functions within the circuit. These algorithms are usually designed (and the functions chosen) using a statistical approach to reveal as much about the functioning of the circuit as possible. This way, it is possible to test a circuit without producing every possible output.

Extremely complex circuits may have to be *designed for testability*, particularly if they are very new designs. The ability to be tested is designed into the circuit. Frequently, extra *test points* are included inside the circuit. These may be used to reveal much about the functionality of the circuit. Fine probes are used to make contact to subsections of the integrated system so that design errors may be tracked down more readily. The test points may even be enhanced by internal *test circuits* which help to *debug* designs after they have been processed by using *self-test* techniques.

Due to the complexity of testing at speed, computers are employed to control the test systems (see the next section). A *test program* is therefore necessary to control the computers. The design of the test program is typically done when the circuit is being designed. This way, the circuit may be designed with testability in mind.

Test Methods

Testing of an IC begins before the circuit is even finished. *Parametric testing*, the testing of parameters relating to the wafer itself or to devices, as opposed to overall circuit performance, starts at the beginning of the fabrication process and continues to the completion of the circuit. Each wafer typically has five or so *drop-ins* placed at strategic sites on the wafer, as shown in Fig. 9.4. The drop-ins contain *test structures* or *test areas* which, when tested using *parametric test equipment*, reveal how successful each process step has been in creating the devices. Some in-process parametric measurements are also performed on *test wafers* which are noncircuit (blank) wafers. These are added to the batch for particular process steps and are usually removed after their job is done.

Parametric test equipment includes:

Four point probe. This is a resistance method which determines if diffusions or implants have produced the correct resistance, junction depth, or dopant concentration. It also determines the resistance and thickness of metal layers.

Ellipsometry. This is an optical method which uses a laser to determine the thickness and *refractive index* (an important physical quantity which relates to density) of oxides, nitrides, and other transparent layers.

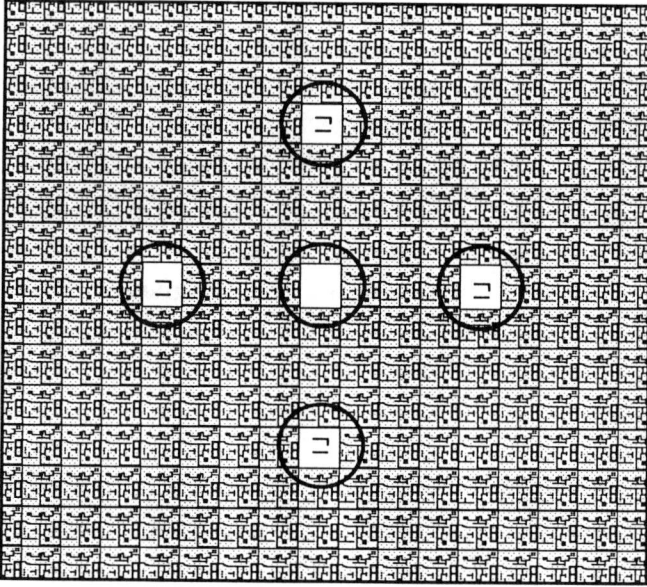

Figure 9.4 Drop-in test sites.

C-V, I-V. These are electrical methods for determining the quality of gate oxides and MOS structure in general. C-V is *capacitance-voltage* analysis, and I-V is *current-voltage* analysis.

If a parametric test reveals a problem during processing, remedial action may be taken. For instance, a *rework* of a certain noncritical stage could be performed (photoresist can be reapplied in many cases). Alternatively, the wafers may be *scrapped* if remedial action is deemed impossible and further processing would be a waste of time and money.

Parametric testing performed at the end of processing is done using drop-ins. This gives full information regarding the results of processing and device parameters. It is design-independent testing, as the test devices are totally separate from the circuit. The drop-ins may be probed and tested using automatic equipment and can provide the necessary device parameters for circuit simulation models.

Functional testing is performed on the finished circuits while they are still in wafer form. It is easier to test relatively small items such as chips when they are in this form. The test equipment is connected to the IC through *probes*, needle-like connections which make contact with the *bonding pads* (see later) and any test pads present. These are held precisely in position by a *probe card* which provides a mechanical base for the probes and electrical connections. To speed up testing with automatic test equipment, an *automatic prober* is used. This steps the array of probes along each IC on the wafer until every circuit is probed and tested. Functional test equipment has to be more complex than the circuit under test. Therefore, the equipment is generally controlled by a large, fast computer which runs complex testing programs.

Functional testing determines whether the circuit performs to design specifications once process parameters have been determined and are found to be correct. Testing can be performed on analog or digital circuits (different test equipment is required for each type).

Surface and Materials Analysis

In addition to parametric and functional testing, ICs during manufacture or at the end of processing may be tested using *surface and materials analysis* methods. These methods measure variables related to the materials used in device manufacture. These parameters involve the type and distribution of impurities, including unwanted contaminants, as well as the integrity of materials and interfaces. This type of testing is not performed on all wafers or circuits but on samples, particularly if problems are suspected.

Typical surface and materials analysis methods are:

Scanning electron microscopy (SEM). This method uses a highly focused beam of electrons to reveal surface topology at extreme magnification.

X-ray crystallography. This method illuminates materials with x-rays and is used to determine crystal quality.

Secondary ion mass spectroscopy (SIMS), auger electron spectroscopy (AES), x-ray photoelectron spectroscopy (XPS). These methods are used to determine elemental content at the surface of materials, as well as the *profiles* or *distributions* of impurities. SIMS bombards the sample with ions and analyzes the ions emitted.   AES hits the material with electrons and analyzes the *Auger* electrons (electrons which come from very close to the surface) emitted. XPS illuminates the sample with x-rays and analyzes the electron emissions. The elemental content may be derived from these analyses.

Energy dispersive x-ray analysis (EDXA). An EDXA system may be added to an SEM to give it materials analysis capability by analyzing the x-rays emitted.

Fourier transform infrared (FTIR). This method uses infrared light to determine, among other things, the oxygen content of silicon and surface conditions.

Diagnostic Methods

In some complex circuits, it is impossible to design in sufficient internal test points. In larger geometry circuits (3 micron or larger), it is possible to use a fine probe to make contact inside the circuits, directly onto the metal interconnects.   This is not possible for multilevel metal circuits or very small geometries.

A probeless or contactless technique called *voltage contrast* is now becoming popular. It uses a beam of electrons in a scanning electron microscope as a probe and measures voltages using the interaction effects of this beam with the conductors in the circuit.   When connected to an appropriate test system, this technique can debug designs by performing measurements within circuits.

One thing that should be kept in mind is that testing of the design once the device has been fabricated should be thought of as a last line of defense. It is an expensive way to find out that there is even a small design fault once the wafers have been produced. Therefore, it is worthwhile to invest time and money in software which will provide as high a degree of device and circuit simulation as possible to find faults before the wafers are made.

Redundancy

In highly complex circuits which use regular arrays of subcircuits (e.g., large memories), the failure of a single component could destroy the functioning of the circuit, even though a million other devices are perfect. To avoid this situation, *redundancy* is employed. Critical components have parallel stand-by elements which will operate if the component fails. This concept may be carried over to entire circuit elements, consisting of thousands of devices, which are duplicated on chip to prevent total failure if critical devices are nonfunctional. Modern ultra-large-scale circuits have the ability to switch function to alternative circuits during operation in the event of device or interconnect failure.

As mentioned previously, some complex circuits have now been designed which can perform a certain degree of *self-testing*. On-board circuits exercise parts of the main circuit when put into *test mode* either during functional testing or during normal operation. The outputs from these tests may be used to activate redundant elements or to signal a fault condition to the user of the IC.

Summary

The design of advanced ICs is performed to a set of process-dependent design rules on computer-aided design systems which utilize a wide range of software packages for synthesis and analysis. Synthesis programs are utilized to aid the designers with the circuit layout, and analysis programs are used for design rule checking and circuit simulation. The most exacting and expensive design process is full custom, but semicustom design provides a faster turnaround for circuits which do not have to be totally optimized. Testing of the circuits is also performed in many cases by computer-controlled systems and falls into two categories: parametric and functional. The former is largely circuit independent, being performed on layers and test structures during and at the end of processing. Functional testing is generally carried out on all product circuits. Surface analysis methods may also be used on samples to determine material properties. Internal test points, self-test capability, and advanced diagnostics such as voltage contrast are used to debug complex designs after they have been fabricated. Redundancy may also be built into complex circuits to allow them to be fault tolerant.

10

Packaging and Interconnections

Interconnection Within ICs

ICs have internal *interconnections* between devices so that they may interact to perform some desired function. The areas of the circuit which contain devices are called the *active areas*, whereas the interconnection regions are called the *field areas*. In modern ICs based on silicon MOSFET technology, the active and field areas are typically delineated by a *local oxidation of silicon (LOCOS)* process. Silicon nitride is used to prevent oxidation over the active areas, whereas a thick field oxide grows in the field areas. This thick oxide supports the interconnection but does not allow accidental device formation. As mentioned in previous chapters, when we integrate devices, we must always be aware of the possible formation of parasitic structures, as these can often severely affect circuit performance.

It is now common to use multiple levels of interconnect within ICs. In modern silicon gate circuits, we have the ability to run interconnection in at least three media: the *substrate diffusion* (formed during the diffusion or implantation of the heavily doped source-drain regions and actually an extension of these), the doped poly-Si, and the final level of metallization. Two levels of metal (often called *dual level metal* or DLM technology) are now commonly found in advanced ICs. Some highly specialized circuits can have even more levels of upper interconnect. More levels of interconnect allow greater degrees of freedom in terms of how the circuit may be configured, and can lead to higher packing densities and greater functionality. The addition of further layers of metal is not as easy as it may seem, but there are many schemes to choose from.

Once the first level of metal has been deposited and etched, an *interlevel dielectric (ILD)* is put on. This must be a good insulator, but it must also *planarize* the surface so that it is smooth enough for the second level of metal. This is important, as a surface with relatively large steps can lead to very poor *step coverage* of the second layer of metal, which in turn can lead to circuit failure. (Metal deposited by physical vapor deposition can thin severely at the bottom of steep steps.) If the first level of metal is aluminum, the ILD should also be deposited at low temperature so as not to destroy the metal (the problem is not as serious if tungsten is used for metal 1, as it can withstand high temperatures). There are a number of ILD schemes which fulfill these requirements. The simplest ones use materials which may be spun on the surface. This is a *self-leveling* process, that is, it leads to a relatively flat layer even if the underlying topography is severe. Materials used in this process are *polyimide*, an organic plastic-like substance, and *spin-on glass (SOG)*. The latter is a soluble compound of silicon dioxide carried in a solvent so that it may be applied in liquid form. After being spun on, the layer is baked at relatively low temperature to leave SiO_2. The alternatives to the spin-on ILDs are *bias-sputtered quartz (BSQ)* and *resist etch-back (REB)*. BSQ involves sputtering the surface during deposition so that any peaks are leveled out. REB relies on photoresist being spun on to a chemical vapor-deposited oxide layer so that it is thin at peaks in the underlying layer. The substrate is then dry etched and the thin resist etches away first, allowing the peaks in the oxide layer to be etched down, thereby planarizing the surface.

Once the ILD is formed, connection holes called *vias* are etched through so that the second level of metal can connect to the first. The second level is then deposited and patterned. If different materials are used for the metal 1 and metal 2 layers, a *barrier layer* may be necessary to prevent unwanted reactions. The last metal layer is almost always aluminum. Even after this is put on, the circuit is still not complete. The last step is to deposit a protective layer of low-temperature chemical vapor-deposited oxide or plasma-enhanced

CVD silicon nitride. This step, called the *passivation*, protects the circuit from moisture and other contamination and from mechanical damage such as scratches during the *packaging* process. If a deposited oxide is used, it is usually doped with phosphorus, which helps to stop *mobile ionic contamination* such as *sodium* from going through and helps to reduce the possibility of cracking. Nitride is highly impermeable and does not need additives, but the layer can contain internal *stress*, which may lead to cracking if the deposition conditions are not ideal.

Packaging Concepts

Once the circuits have been tested in wafer form and the nonworking ones identified (usually by an ink dot), the individual ICs have to be separated and the good ones *packaged* so that they may be used in electronic systems. Packaging allows electrical connection of the circuit to the outside world. This includes signal or data-in connections, signal or data-out connections, and control and power lines. Packaging also provides protection for the delicate ICs in harsh environments. Environmental hazards include moisture, temperature, corrosion, vibration, and other mechanical shocks. Finally, packaging provides a way for heat generated in the circuit to be dissipated so that the IC does not overheat.

Packaging is an important area in integrated electronics. The original packages for discrete devices were very simple. They were basically metal cans which covered a base to which the semiconductor was attached. The base had long metal wires or leads protruding to allow connection of the device to a printed circuit board. As we will see later in this chapter, packaging has become much more complex to meet the needs of more sophisticated electronic systems.

Dicing

To separate the individual ICs on the wafer after functional test, the wafer must be *saw cut* or *scribed*. This process is called *dicing*.

When the circuit is designed, a plain border, called the *scribe channel*, is put on . This allows the wafer to be broken up without damaging the edges of the circuits. Scribing uses a diamond-tipped scribe to "score" the wafer along the scribe channels. The wafers may then be broken to separate the individual ICs (the process is quite similar to glass cutting). Saw cutting is much more precise, as it involves deep saw cuts into the scribe channels and does not rely so much on a "clean break" as the scribing technique. A diamond-edged saw is used for this purpose. The wafer is placed on an automated table which indexes the wafer one column of chips at a time and then rotates through 90 degrees to cut the rows. After dicing, the marked failed devices are discarded or perhaps sent for *failure analysis*. The good chips then go on to the next stage.

Mounting and Bonding

The working ICs are gathered and *mounted* in an appropriate package. Mounting an IC involves attaching it to the base of the package so that a good thermal and electrical contact is made between the chip and the metal base. This is generally achieved through the use of an electrically and thermally conducting *silver epoxy resin* or a *gold eutectic bond*. The latter forms an excellent electrical, thermal, and mechanical bond, as it relies on the fact that gold and silicon will melt together and form an alloy at a temperature (called the *eutectic temperature*) which is much lower than the melting points of either material in isolation. Immediately prior to dicing, the wafer is *back lapped*, a mechanical and chemical means of thinning the wafer to improve the electrical contact to the devices, and gold is evaporated onto the back. During heating of the chip on the package base, the gold and silicon mix and melt onto the package, which has a gold-coated chip mount, at around 425° C—the eutectic temperature for these materials.

When the ICs are designed, the connections which have to link the internal portions of the circuit to the outside world are led to *input/output* (I/O) structures. These I/O units contain appropriate *buffers* and *drive circuitry* to handle the input and output voltages and currents, respectively, and prevent overloading; *static protection* structures to prevent damage from static electricity; and *bonding pads*. For instance, in digital circuits, the input buffers may just be single transistors, whereas in analog circuits they may be FET input operational amplifiers. Also, in a BiCMOS microcontroller, the output drivers will almost certainly be large BJTs. The static protection structures range from simple *spark gaps* to sophisticated *gated diodes*. The bonding pads enable connections to be made between the circuit and the package.

After the IC is mounted, connections are made to the package by a process called *bonding*. This involves a high-precision machine which bonds a fine (about 50 micron) gold or aluminum wire onto the bonding pad and connects the other end to the appropriate connection on the package. The process is shown in Fig. 10.1. The bond wire is passed through a *capillary* and is heated by a spark to form a ball at the end. The ball is pressed onto a bonding pad and is attached to it by heat or *ultrasonic* (high-frequency sound) energy. The capillary is then moved to the connection on the package, allowing the wire to travel with it, and pressed onto this, forming a compression bond. The wire is locked and the capillary raised so that the wire breaks, leaving an end which may be balled for the next bond.

This process was traditionally performed under operator control and was very labor intensive (many ICs require 40 bonds or more per circuit). Therefore the operation was performed in countries where labor was inexpensive. Now, computer-controlled bonders need little operator intervention.

Tape automated bonding (TAB) is used for small-size, high-volume ICs to save on bonding cost and time. The circuits are first *bumped*, that is, the bonding pads have gold bumps applied. These bumps may then be connected to *prefabricated copper fingers* by means of a *thermocompression* (heat and pressure) or *eutectic* bond. This is a quick and inexpensive method, as the connection preforms or leadframes may be run as a continuous *tape*; therefore high-speed automation is possible. Also, all bonds to the chip are made simultaneously when a heated block presses the fingers onto the bumps. This is faster and cheaper than wire bonding.

Package Types

As mentioned previously, packaging for discrete transistors is relatively simple, as there are few connections to be made from the device to the external circuit (usually three for a transistor). Traditionally, discrete transistors were packaged almost exclusively inside miniature metal cans, but a molded plastic bead is now a more popular package type for small-signal devices. ICs can have many tens of connections, and therefore the package has to be more complex. There are many different types of IC package. The main ones are described in this section.

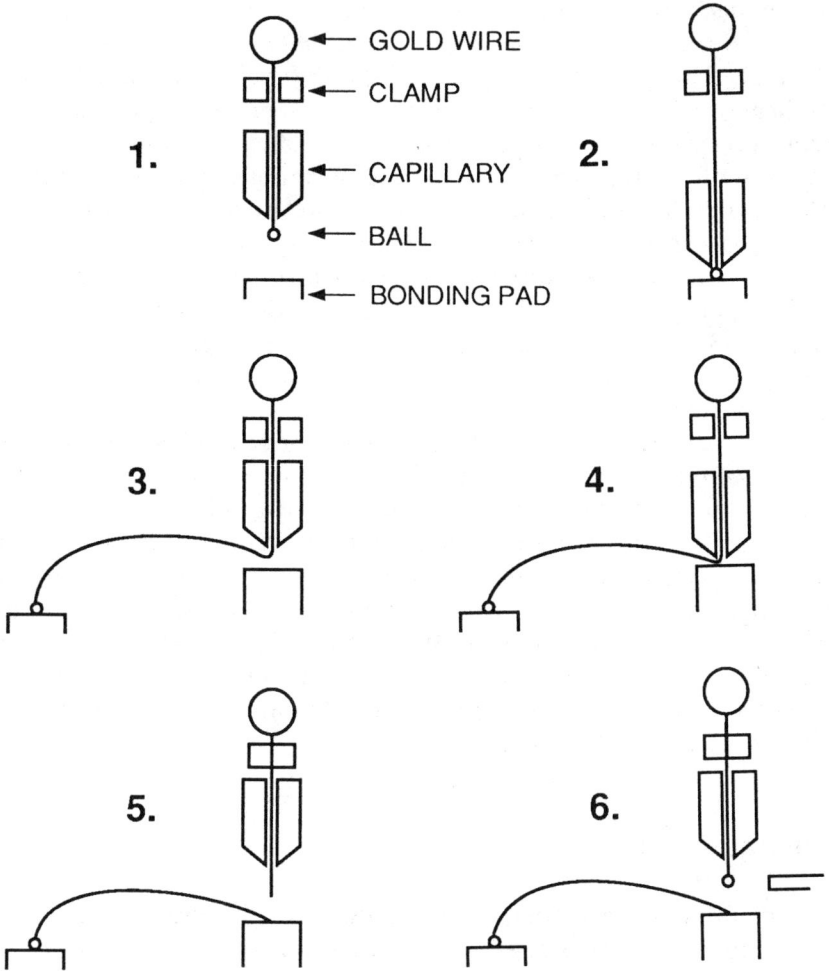

Figure 10.1 Ball bonding technique.

A *dual in-line package (DIL or DIP)* is a molded epoxy plastic package formed around a bonded circuit (Fig. 10.2). The plastic is melted and molded around the wire-bonded circuit on its leadframe to form a good seal. The *pins* are the connections which will ultimately be attached to the copper tracks of the printed circuit board, either directly or via an *IC socket*. A good seal is necessary to prevent the environment from damaging the IC or the bonds. Plastic DIP is a low-cost, high-volume packaging method. DIP may also use *ceramic* material, *aluminum oxide* (alumina). This is a more expensive material and is harder to work with, but it has better heat conduction properties and is more robust than plastic. (Plastic has another curious disadvantage: the material tends to contain minute amounts of radioactive elements, and the radiation can affect memory circuits.) The ceramic material is fed into a mold as a slurry and baked hard around the leads. The final package is made with a space for the circuit to rest in.

The DIP has the connections to the outside world along the long edges of the package, and comes in standard pin pitches and standard

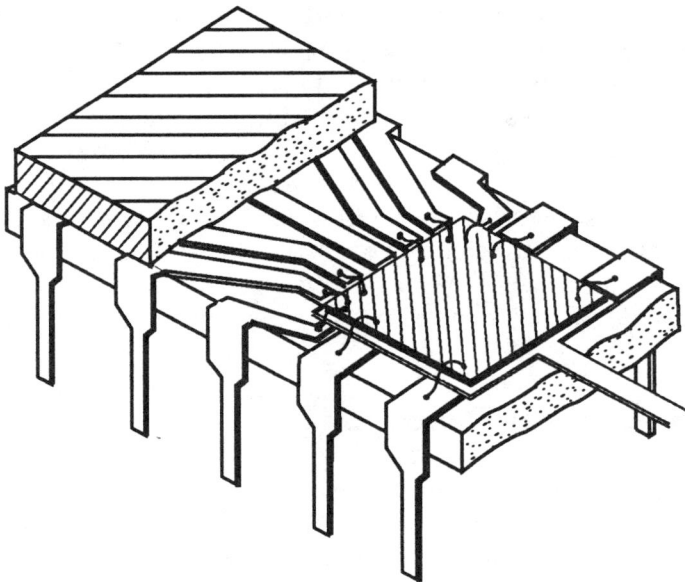

Figure 10.2 DIL package (cut away).

numbers of pins (e.g., 8, 16, 24, and 40 pin, half the number of pins per side of the package in each case). High thermal performance packages for low-volume, high-cost applications tend to be *ceramic DIP (CERDIP)* types. CERDIP uses a preformed ceramic package with the package connections already in place. The package may be in two halves (split down the plane of the IC mounting pad) or may have a small lid. The seal between the halves is generally a low melting point glass, whereas the lids are attached and sealed by metal (solder). These packages make an excellent hermetic seal; this is an absolute seal which excludes moisture and gas diffusion to the inside. Note that CERDIP is better for high-reliability applications, such as circuits for military or aerospace applications as the material is thermally and mechanically stable.

DIP has the disadvantage that increasing the pin number beyond 60 creates a very long package. Therefore other packaging technologies are necessary for ICs with large numbers of pins. *Chip carriers* are similar to DIP, except that they tend to be square and have connections or leads on all four sides. Beyond chip carriers are *pin grid array (PGA)* packages. These are complex packages which allow connections through a regular square array of pins, as shown in Fig. 10.3. These packages may have hundreds of pins. The connections from the IC to the pin grid package usually have to be made via *flip chip* technology, as conventional bonding from the edge of the chip to the package would result in overcrowding of the bonds.

Flip chip technology involves the *bumping* of the bond pads of the IC just as in TAB. However, in this case, the circuit is turned face down on a preformed leadframe in the PGA package. The package is heated and a eutectic bond is formed between the bumps and the leadframe. This is fast, as all the bonds are made simultaneously and bonding pads can be placed virtually anywhere on the circuit.

Current Trends in Packaging

Some computer manufactures are now using *multilevel* systems which interconnect several packages together into one unit. This approach is typically used for fast processors in mainframe computers. It can lead to higher reliability, as high-performance ceramic substrates are used which dissipate heat and are more stable mechanically. The unit generally has to have a special cooling system, but this is no problem for nonportable systems.

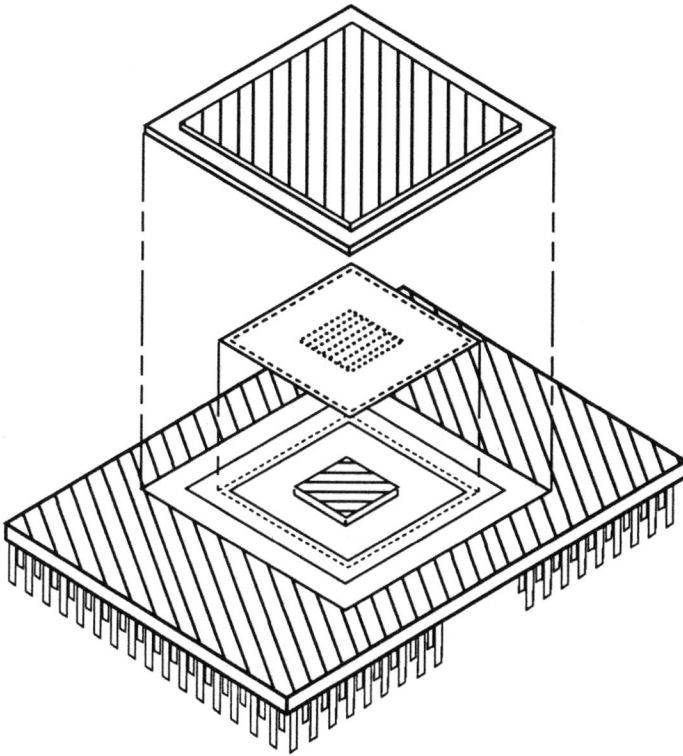

Figure 10.3 Pin grid array package.

Surface mount technology (SMT) is now becoming an alternative to packages with leads in some applications. Traditional DIPs, chip carriers, and PGAs have metal leads or pins which would normally be inserted through holes in printed circuit boards and soldered in place to the copper tracks. Surface mount packages can also be plastic or ceramic but have no wire leads or pins. Instead they have large versions of the bumps used in flip chip technology. These big bumps are composed of various layers of metals and alloys. The chips are placed on the surface of the wiring board and heated. The heating melts the bumps, forming the electrical and mechanical connection between the packages and the tracks on the circuit board.

SMT is easier to use for many applications, as it reduces the need for a high-precision automatic component placement assembly system to put leads through small holes in a printed wiring board (these systems fail miserably if any of the leads are bent during handling). It is also more compact, as there are no large packages taking up space on the wiring board; the package in SMT is not much larger than the chip itself. We therefore find SMT in many highly portable systems such as miniature headphone radios and pocket-sized cassette players. The technique is also generally more reliable than the standard drilled PWB approach, as there is less chance of a poor joint being formed. This is because the connecting alloys which take the place of solder are already at the joint and do not have to flow into a small hole to complete the connection.

Perhaps the ultimate in chip-to-chip connection is the *multichip* module. Rather than having individual chips in their individual packages, multichip modules have several ICs in the one superpackage. This has advantages with regard to reliability and in the assembly of circuit boards, as it is easier to place and connect one multichip module than several separate ICs.

Thermal Problems

When high-density circuits are operated at high speed (e.g., ECL circuits in large computers), a great deal of energy is dissipated, manifesting itself as heat. This heat could destroy a circuit through *thermal runaway*, a form of self-destruction, and therefore must be channeled away from the circuit.

It is, of course, important to ensure that the heat flows away from the circuit. That is why the chip is mounted using silver epoxy or a eutectic bond on a metal base. This situation provides high *thermal conductivity*. The heat from the IC thereby flows into the package. Ceramic materials can dissipate heat reasonably well and remain intact even at very high temperatures. Plastic packages will crack if overheated.

Often *heat sinks* are used on the packages to remove heat by convection. These are finned metal structures attached to the package using a bolt or clamp arrangement. Alternatively, the package may simply be bolted to the chassis or case of the system. For circuits which can get extremely hot, forced cooling may be necessary. Large modern computer ICs use a refrigerant coolant in an effort to keep temperatures down. For multilevel packages, Freon or another coolant can actually be fed into and through the package.

Hybrid Circuits

Not all ICs are destined to be directly packaged after completion. Some may be used in *hybrid* circuits. Hybrid circuits are based on a *ceramic substrate*. The substrate has high-precision resistors put on by a *thick film* technique, essentially screen printing, using a glass and metal mix. These resistors may be *laser trimmed* to high accuracy. High-value capacitors may also be included. *Chip capacitors* are single large, leadless capacitors which may be mounted directly on the substrate, much in the same way as SMT components. ICs are then mounted and connected by wire bonding directly to the substrate. The entire hybrid circuit substrate may then be packaged in a metal can or in epoxy.

One of the main advantages of hybrid technology is the high precision of the trimmed substrate resistors external to the ICs. Such high precision is useful in, for example, measurement applications. This degree of precision is very difficult to achieve in diffused on-chip resistors. Also, hybrids can be used in high-voltage applications such as telephone switching. The voltages can be reduced by the circuitry on the substrate so that damage to the onboard ICs does not occur. Since the ICs are mounted on a ceramic substrate, thermal problems are also reduced.

Wafer Scale Integration

An emerging technology in high capacity memory systems is *wafer scale integration (WSI)*. WSI uses an entire wafer of similar circuits connected together. There are no scribe channels, so that the metal tracks may interconnect every circuit on the wafer. Redundancy techniques are employed, as are self-tests of the individual units. Each good unit tests its neighbor for full functionality and routes the signals through the good elements. The failed units are permanently switched off. This way, extremely large memory capacities are possible and system yields are high. The wafer is never diced but is packaged whole in a very large carrier.

Summary

The interconnections between the individual devices within an IC are attained via diffusions, poly-Si layers, and multiple levels of metal.

However, in order for IC to be connected to an external circuit, it has to be put in an appropriate package. The completed and tested wafer is diced and the working chips are mounted so that the are electrically, mechanically, and thermally attached to the package. Connections are made between the IC and the pins of the package by means of fine bonding wires or preformed leadframes. The two main package materials are plastic, which is inexpensive, and high-performance ceramic. The style may be DIP, chip carrier, or PGA. Current packaging trends are moving towards SMT and multichip packages. Hybrid circuits, which mix ICs and thick film technology, are also used in applications which demand high-precision resistors. Many high-current or high-speed circuits generate so much heat that heat sinking or forced cooling may be necessary.

11

Cleanrooms, Yield, and Reliability

The Concept of Yield

For a circuit manufactured using a particular technology to be successful, it must be capable of being produced in large quantities at costs which are competitive with those of alternative methods. The circuit must also be capable of performing its function throughout its intended lifetime. To produce circuits which meet these criteria, a knowledge of why high costs and unreliable devices occur is necessary. This is the realm of *yield* and *reliability*. Yield is the proportion of circuits which are functional, expressed as a percentage of the total possible working devices; for instance, if we have 200 integrated circuits on a wafer and only 100 work when tested, the yield is 50 percent. Reliability is the quality of a circuit to survive beyond its infancy stage and continue to operate for some time afterward.

Some legitimate questions may be asked at this point. Why should any circuits be declared dead or dying when they are tested immediately after fabrication? In other words, why should the yield not be 100 percent? Why should they fail after some time in service? The answers to these questions are rather complicated, but we shall examine the main points in this chapter.

Failure Modes—Yield

In practice, the yield may be close to 100 percent (all devices working) or close to zero. There are essentially three categories of causes for yield to be less than 100 percent:
1. Parametric processing problems
2. Circuit design problems
3. Random point defects

IC fabrication techniques are not ideal and cannot always be controlled absolutely. Therefore, variation of process parameters is not uncommon. Process variations may lead to the nonfunctioning of circuits if they occur in critical areas. Process variations include:

Variations in the *thickness* of grown or deposited layers (gate oxide, poly-Si, etc.)

Variations in the *resistance* of diffused or implanted layers (source-drain diffusions, threshold adjust implants, etc.)

Variations in the *width* of lithographically defined lines and features

Variations in the *registration* of a photomask with respect to previously defined layers

All of these variations could render a circuit inoperative if they fall outside of the *tolerance* limits. For instance, if the gate oxide in a MOS device is too thin, then the device characteristics will be severely altered (the threshold voltage in particular will change) and the device may even break down when a voltage is applied to the gate. Variation in a process parameter can also be uneven across a wafer. It is possible for one-half of a wafer to contain perfect devices, while none of the devices on the other half work.

Poor *circuit design* can lead to sensitivities of the circuit to combinations of process parameters which are within the tolerance of the process. This can happen if too little thought has been given to the correlation between variations in different process parameters. For instance, the circuit may fail due to the worst case combination of process variations.

Point defects are small regions of imperfection, for smaller than the size of the circuit. These may be caused by slight imperfections in the crystal substrate formed when the crystal is grown or cut, or may be created when the substrate is heated or mishandled. A further and usually more significant source of point defects is *particulate material*. Tiny particles in the air from outside sources, operators, construction materials, process materials, and the processing equipment can create defects. The circuit defects may form during photolithography when the particle acts as a mask and its image is transferred to the layer on the wafer. They may scratch the wafer during photolithography. They may be burned in during a furnace process and chemically contaminate the wafer. Finally, they may disrupt the continuity of a thin deposited layer by causing *pinholes*.

Particulates can range from a few tenths of a millimeter to fractions of a micron. Killer particles tend to be those which impinge on sensitive active areas and which are greater than one-tenth of the minimum linewidth. Larger-area circuits naturally mean reduced yield, as there is more chance of a killer defect existing somewhere

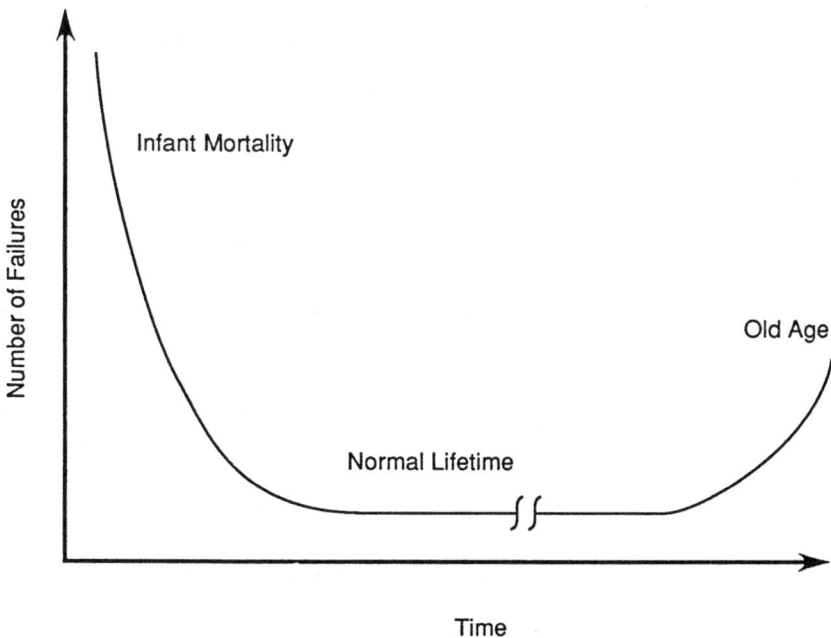

Figure 11.1 Bathtub curve.

within the circuit. Put another way, if a number of point defects are spread out across a wafer, a larger circuit is more likely to encounter one of them. Designers therefore try to keep the circuit area as small as possible to lessen this effect.

Failure Modes—Reliability

A process problem need not render a circuit inoperative immediately. It may take hours or years before any effect is noticed. There are also other processes which can cause a circuit to fail in operation. The number of failures vs. time is usually represented by the so-called *bathtub curve* of Fig. 11.1. The high rate of *infant mortality* is due to processing and assembly problems. The large number of *old age* victims is due mainly to *wearout*. However, a portion of the devices may fail in the intervening period.

Time-dependent failure mechanisms include:

Contamination, for example, diffusion of mobile alkali metal ions into the substrate or gate oxide

Electromigration, the thinning of metal tracks particularly at weak points under current flow, causing conductor failure

Corrosion of interconnects and bonds by ingress of chemicals into leaky packages or by residual contamination

Growth of unwanted compounds at contact points between metals

All of these processes are accelerated by increasing the temperature. Other failure mechanisms include:

Electrostatic damage (static electricity) which disrupts the gate oxide

Mechanical shock (vibration) which disrupts bonds

Radiation damage which creates transient errors in MOS memories

Thus there are many reasons why both yield and reliability may be reduced. The mechanisms are fairly well understood but still tend to be somewhat unpredictable. Therefore they inevitably lead to problems.

Burn-in and Accelerated Testing

Since temperature accelerates failure processes, devices may be preconditioned to screen out early failures by heating. The process is called burn-in. Any devices which are likely to fail due to process faults should fail during burn-in and thus will not find their way to the marketplace.

To determine the lifetime of devices which have passed burn-in requires *accelerated testing*. Most of the time-dependent failure modes may be brought on by accelerated testing by increased *temperature* and *temperature cycling*, increased *voltage* and *current*, and increased *humidity*. The packaged ICs are loaded into sockets on wiring boards, which are loaded into large ovens providing the appropriate environment. Severe *vibration* may also be utilized. Such accelerated testing can give information on *mean time between failures (MTBF)*. This is basically the average expected lifetime of a particular product.

Specifications

Circuits may be designed, fabricated, and tested to various *specifications*. Manufacturers frequently offer devices, which may actually be functionally identical, which fall into different specification categories. These categories typically are determined by temperature, mechanical stability (usually a function of the packaging), and occasionally operating parameters such as speed.

A *commercial* specification may have an operating temperature range of -5° C to 85° C, whereas a *military* specification may be -50° C to 150° C. The commercial package may be plastic, whereas the military package is typically ceramic. Circuits destined for military and space applications have to undergo full burn-in and strict testing to agency guidelines. A full range of accelerated testing may have to be performed on large representative samples so that the manufacturers are certain that the devices will meet military specifications.

Quality Control

One of the most critical engineering functions in IC manufacture is *quality control (QC)* or *quality assurance (QA)*. QC ensures that all process instructions are followed absolutely to minimize process errors. QC departments perform parametric tests on work-in-process to ensure that quality criteria are being met.

It is generally the job of a QC department to monitor accelerated lifetime tests and to correlate data with the results of processing, operator habits, etc. With the advent of techniques such as *statistical process control (SPC)*, QC is now much more of a science in semiconductor processing. SPC uses statistical mathematics to ensure the validity of data and data collection methods. *Control charts* are usually used in production processes to record a parameter sequentially (e.g., the thickness of an oxide on a batch-by-batch basis). If the parameter exceeds a predetermined *control limit* on the chart, remedial action may be taken to bring the process back under control.

Controlled Environments

The production of ICs is a very precise and delicate operation. All process steps must be controlled absolutely, which means that the *fabrication environment* must also be controlled. The main parameters which must be controlled in a processing environment are:

- Particles
- Temperature
- Humidity
- Room pressure
- Air velocity and flow
- Vibration
- Static electricity

Control of these parameters may be achieved by constructing a specialized room, called (for obvious reasons) a *cleanroom*, and supplying conditioned air to this room.

Particulates (dust) may be controlled by *high-efficiency particulate air (HEPA)* filters, temperature may be controlled by heating and cooling coils in the airstream, and humidity can be controlled by cooling the air to remove moisture or spraying water to increase moisture content. A typical *air handler* system is shown in Fig. 11.2. Most of the air which is conditioned and put into the cleanroom is actually recirculated in the system, as conditioning consumes energy and is therefore expensive. Fans and dampers control the air velocity and room pressure, and vibration is controlled by isolating the sources or the equipment.

The construction materials, the processing equipment, and even the operator apparel used in the cleanroom must be compatible with the clean environment. Systems supplying the cleanroom with, for instance, water and gases should also be compatible with the controlled environment concept. Consequently, process gases are supplied via high-quality stainless steel lines, and water and other liquids are brought in through pipes made of inert plastics.

Microcontamination

The natural environment is a source of contamination which is detrimental to IC processing. Particulates come from a variety of sources including the earth, construction materials, plants, and humans. The sizes of particles carried in the air range from about 100 microns down to hundredths of a micron. In a typical air sample, the number of particles a few tenths of a micron in size number between a few hundred thousand and a few million per cubic foot.

Particles measuring 0.3 micron may be removed to 99.9999 percent efficiency by HEPA filters. *Federal Standard 209* defines classes of cleanrooms with regard to particles carried in the air. For instance:

Class 100,000 means that there should be no more than 100,000 particles per cubic foot measuring 0.5 micron.

Class 10,000 means that there should be no more than 10,000 particles per cubic foot measuring 0.5 micron.

Class 1,000 means that there should be no more than 1,000 particles per cubic foot measuring 0.5 micron.

Class 100 means that there should be no more than 100 particles per cubic foot measuring 0.5 micron.

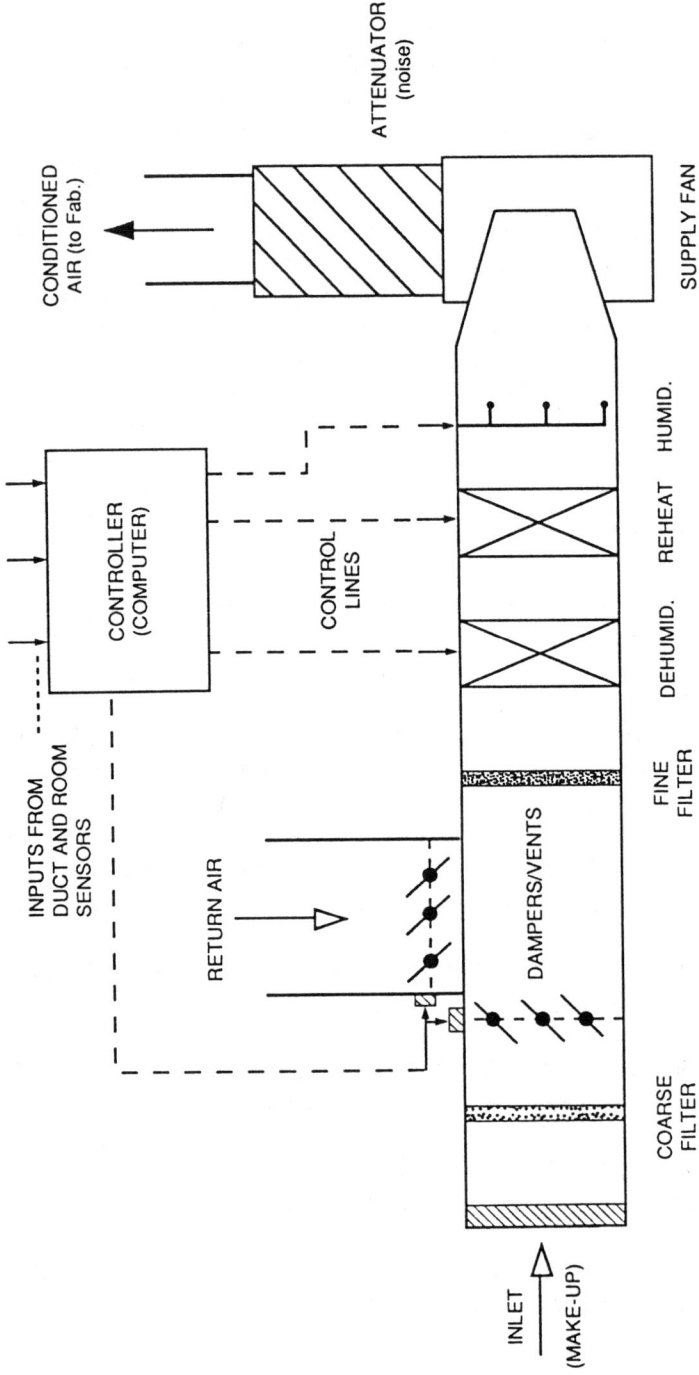

Make-up air is typically 20% of total volume
(to provide fresh air and compensate for losses from doors, hoods, etc.)

Figure 11.2 Typical air handler for conditioning of cleanroom air

Class 10 means that there should be no more than 10 particles per cubic foot measuring 0.5 micron.

Class 1 means that there should be no more than 1 particle per cubic foot measuring 0.5 micron, a very tall order indeed, considering that the outside air may contain millions of particles of this size.

Work areas in microelectronics are typically class 100, 10, or better, whereas service and assembly areas are about class 10,000 or 1,000.

To maintain a clean environment over equipment and work surfaces, the air is made to flow through the HEPA filters at about 100 feet per minute, creating a nonturbulent *unidirectional* or *laminar flow*. Airflow of this type of clean air helps keep the work areas clean by sweeping away any particles which enter the area from, for example, equipment or operators. Much of the air is then recirculated to minimize the cost of conditioning and filters.

Since a large source of particles is human operators, personnel should be appropriately attired to prevent particles on the body or clothing from reaching the product. For a class 100 environment, this means that the operator has to wear a *bunny suit* (a full one-piece suit), a hood, boots over outdoor footwear, gloves, and a face mask. All suit materials are cleanroom compatible, For instance, a typical suit material will not allow particles to pass through to the outside or to generate particles itself. Materials which are used are tightly woven *polyester* and *expanded PTFE laminate*.

Cleanroom Systems

All process materials entering the cleanroom must be treated to be fully compatible with the cleanroom environment. In addition, many materials which are used in the production process are hazardous and must be appropriately contained, monitored, and treated before being discharged to the external environment.

The main systems which are used in controlled environments in addition to the air handling units are as follows:

Ultrapure water. Raw water is treated by *reverse osmosis (RO)* and *deionization (DI)* to remove chemicals, including dissolved mineral salts. It is also *filtered* to 0.1 micron or better to remove particulates and bacteria. Ultrapure water is absolutely necessary, as the product is rinsed frequently during chemical etching and cleaning operations. A schematic diagram of a typical ultrapure water system is shown in Fig. 11.3. The water is first pretreated to remove the large-scale material and then passes through the RO units. RO is basically a molecular filter. The water is than further purified by the DI units, which remove dissolved salts by chemical means. The water is then put into the recirculation loop, where it is repeatedly deionized, filtered, and irradiated with ultraviolet light to kill any remaining bacteria. It is kept flowing through the distribution system to prevent stagnation. The water quality is often expressed in terms of its *resistivity*, and the best is rated at *18 megohm-centimeter.*

Gases. *Process gases* are filtered to 0.1 micron or better to remove particles. Gas purity is maintained by the manufacturer of the gases, and clean handling systems using stainless steel piping are used within the semiconductor production plant. Gases which are used in large quantities, such as *nitrogen* and *oxygen,* are stored in liquid form to save space; *liquid nitrogen* tanks are often seen outside semiconductor plants. Special process gases such as *silane, phosphine,* and *ammonia.* are stored in cylinders. These gases are dangerous, and therefore have to be stored and used in safe cabinets and monitored for leaks.

Waste. *Reaction by-products* and waste gases are *burned* and *scrubbed.* They are collected by *exhaust systems* which are connected to chemical benches or *hoods,* gas cabinets, and equipment. Waste *acids* are gathered by a drain system to be *neutralized* and waste *solvents* are collected and recycled or disposed of by *burning* or *burial.*

Of course, semiconductor facilities must have other systems which all factories are required to have, such as fire suppression systems.

Circuit Dimensions

During manufacture, ICs are very sensitive to the environment. In addition, production materials such as photoresists and diffusion sources are very sensitive to temperature and humidity. More

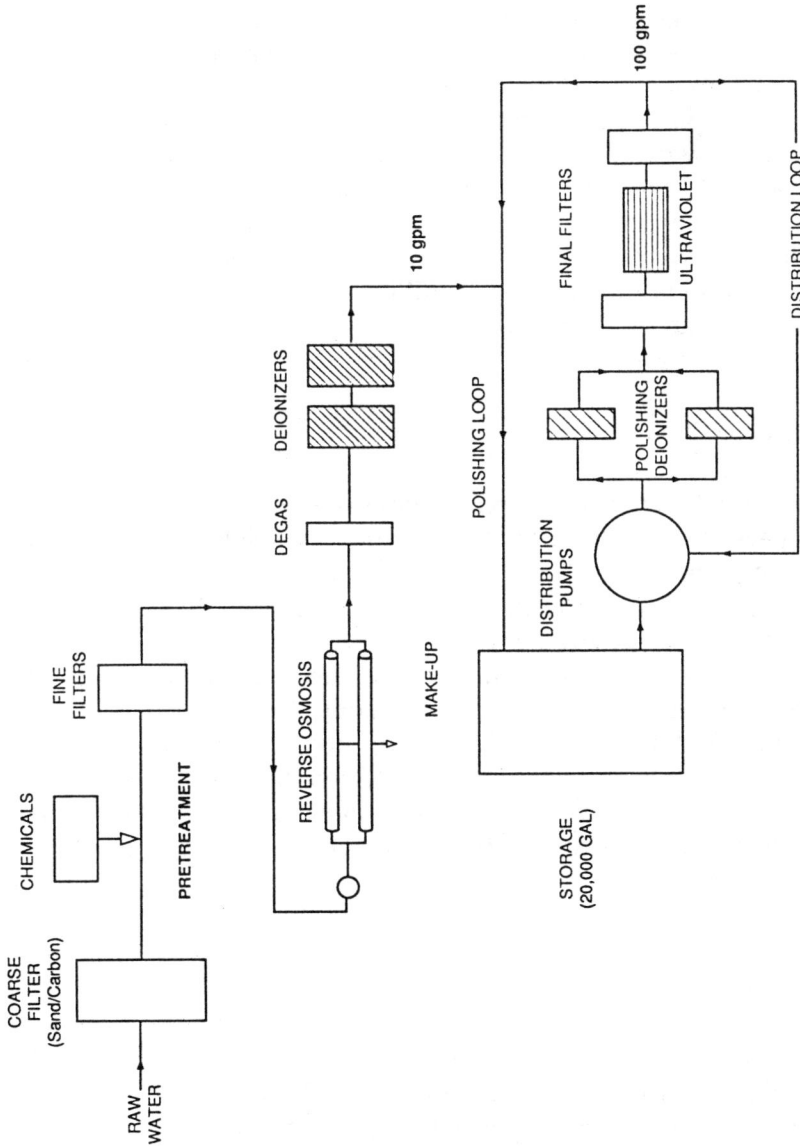

Figure 11.3 System for the production of ultrapure water.

importantly, as we have seen, particulate contamination can disrupt devices within an IC during photolithography or during a high-temperature operation, seriously affecting yield and reliability.

As linewidths in advanced ICs decrease steadily below 1 micron to meet the needs of packing density and speed, particulates become increasingly important. A 0.1 micron particle can seriously affect the performance of a 1 micron linewidth circuit. Therefore more emphasis is being put on control of particulates through better filtering and more protective apparel for the operators. To create a class 1 environment requires a view of the cleanroom as a total system. Every part of the system, from air filtration to gas delivery, is a potential weak link in the chain. Therefore, the design, construction, and operation of cleanrooms has become a science, well beyond the technology of more traditional production plants. Unfortunately, the total clean approach is expensive, with large facilities costing as much as $500 million to build and millions of dollars per year in utility costs alone.

Class 1 and Beyond

One way of achieving a superclean environment is to remove the human operators. This may be done by fully enclosing the process or by replacing the operator by *robots*. Production lines which use mechanical handling for critical processes are already in use but require massive capital investment.

Clean transfer tunnels are also being used in conjunction with automatic loading stations, but the best form of process enclosure is the *standard mechanical interface (SMIF)* approach. SMIF uses sealed boxes for wafer transfer around the production area. These boxes mate with mechanical interfaces on equipment. Therefore the wafers need never see the external environment, which may be class 10,000 rather than class 10.

Summary

There are many reasons why we may not achieve 100 percent yield or total reliability in a semiconductor product. Processing problems and contamination, including particulate materials which create point defects, can render ICs inoperative either immediately or after some time of operation. We can take circuits past the infant mortality point by burn-in, and accelerated testing can give information on how long the product will survive on average.   Products which have demanding specifications for military or aerospace use must undergo more strict burn-in and testing. To control point defects which can occur during manufacture, ICs must be fabricated in controlled environments. Cleanrooms are manufacturing systems in which the air and production materials are cleaned before being put in. Even the operators are gowned to prevent their contamination from reaching the product. There is currently a move away from expensive and expansive cleanrooms toward enclosed systems which surround the products.

Further Reading

Due to the introductory nature of this book, we have only really scratched the surface of the topic of modern electronics. If you wish to delve more deeply into the subject, there are literally hundreds of more advanced texts available, far too many to list here. However, I will make mention of a few which have been useful in writing this book and in my teaching.

Physics of materials:
L. Solymar and D. Walsh, *Lectures on the Electrical Properties of Materials*, Oxford University Press, Oxford, 1988.
L.A.A. Warnes, *Electronic Materials*, Van Nostrand Reinhold, New York, 1991.
K.W. Boer, *Survey of Semiconductor Physics*, Van Nostrand Reinhold, New York, 1991.

Device physics:
S.M. Sze, *Physics of Semiconductor Devices*, Wiley, New York, 1981.
B.G. Streetman, *Solid State Electronic Devices*, Prentice Hall, Englewood Cliffs, 1990.
R.S. Muller and T.I. Kamins, *Device Electronics for Integrated Circuits*, Wiley, New York, 1986.

Fabrication:
S. Wolf and R.N. Tauber, *Silicon Processing for the VLSI Era*, Lattice Press, Sunset Beach, 1986.
W. Maly, *Atlas of IC Technologies*, Benjamin Cummings, Menlo Park, 1987.
W.R. Runyan and K.E. Bean, *Semiconductor Integrated Circuit Processing Technology*, Addison-Wesley, Reading, 1990.
G.E. Anner, *Planar Processing Primer*, Van Nostrand Reinhold, New York, 1991.

Circuits:

M.N. Horenstein, *Microelectronic Circuits and Devices*, Prentice Hall, Englewood Cliffs, 1990.

R.J. Smith, *Circuits, Devices and Systems*, Wiley, New York, 1976.

Design:

J.F. Wakerly, *Digital Design Principles and Practices*, Prentice Hall, Englewood Cliffs, 1990.

J. Mavor, M.A. Jack and P.B. Denyer, *Introduction to MOS LSI Design*, Addison-Wesley, Reading, 1983.

C. Mead and L. Conway, *Introduction to VLSI Systems*, Addison-Wesley, Reading, 1980.

Testing:

M. Abramovici, M.A. Breuer and A.D. Friedman, *Digital Systems Testing and Testable Design*, Computer Science Press, New York, 1990.

Packaging:

R.R. Tummala and E.J. Rymaszewski (Eds.), *Microelectronics Packaging Handbook*, Van Nostrand Reinhold, New York, 1991.

Cleanrooms:

M.N. Kozicki, *Cleanrooms: Facilities and Practices*, Van Nostrand Reinhold, New York, 1991.

Index